Angela Dibenedetto and Michele Aresta (Eds.)

Inorganic Micro- and Nanomaterials

Inorganic Micro- and Nanomaterials

Synthesis and Characterization

Edited by
Angela Dibenedetto and Michele Aresta

DE GRUYTER

Editors

Professor
Dr. Michele Aresta
Department of Chemistry
University of Bari
Via Celso Ulpiani 27
70176 Bari
Italy
michele.aresta@uniba.it

Professor
Dr. Angela Dibenedetto
Department of Chemistry
University of Bari
Via Celso Ulpiani 27
70176 Bari
Italy
angela.dibenedetto@uniba.it

ISBN 978-3-11-030666-8
e-ISBN 978-3-11-030687-3

Library of Congress Cataloging-in-Publication data
A CIP catalog record for this book has been applied for at the Library of Congress.

Bibliographic information published by the Deutsche Nationalbibliothek
The Deutsche Nationalbibliothek lists this publication in the Deutsche Nationalbibliografie; detailed bibliographic data are available in the Internet at http://dnb.dnb.de.

Typesetting: PTP-Berlin Protago-TEX-Production GmbH, www.ptp-berlin.de
Printing and binding: Hubert & Co. GmbH & Co. KG, Göttingen
♾ Printed on acid-free paper
Printed in Germany

www.degruyter.com

This book is dedicated to:
Rosangela
Mattia, Nicolò, Federica, Gabriele

List of contributing authors

Giovanni Agostini
Department of Chemistry
NIS Centre of Excellence and
INSTM Reference Center
University of Turin
10125 Torino, Italy
Chapter 5

Angelo Agostino
Department of Chemistry
NIS Centre of Excellence and
INSTM Reference Center
University of Turin
10125 Torino, Italy
Chapter 5

Davide Altamura
Istituto di Cristallografia
Sede di Bari
70126 Bari, Italy
Chapter 6

Angela Altomare
Istituto di Cristallografia
Sede di Bari
70126 Bari, Italy
Chapter 4

Brunella Maria Aresta
Istituto di Cristallografia
Sede di Bari
70126 Bari, Italy
Chapter 6

Michele Aresta
CIRCC
70126 Bari, Italy
Chapter 8

Silvia Bordiga
Department of Chemistry
NIS Centre of Excellence and
INSTM Reference Center
University of Turin
10125 Torino, Italy
Chapter 5

Elisa Borfecchia
Department of Chemistry
NIS Centre of Excellence and
INSTM Reference Center
University of Turin
10125 Torino, Italy
Chapter 5

Dario Brancaleoni
STEPBIO Srl
40133 Bologna, Italy
and
Zinsser Analytic GmbH
Frankfurt, Germany
Chapter 2

Antonella Colucci
CIRCC
70126 Bari, Italy
Chapter 8

Roberto Comparelli
CNR-IPCF Istituto per i Processi Chimici e Fisici
Sez. Bari, c/o Dip. Chimica
70126 Bari, Italy
Chapter 7

Corrado Cuocci
Istituto di Cristallografia
Sede di Bari
70126 Bari, Italy
Chapter 4

M. Lucia Curri
CNR-IPCF Istituto per i Processi Chimici e Fisici
Sez. Bari, c/o Dip. Chimica
70126 Bari, Italy
Chapter 7

Liberato De Caro
Istituto di Cristallografia
Sede di Bari
70126 Bari, Italy
Chapter 6

Angela Dibenedetto
Department of Chemistry
University of Bari
70126 Bari, Italy
and
CIRCC
70126 Bari, Italy
Chapter 8

Elisabetta Fanizza
CNR-IPCF Istituto per i Processi Chimici e Fisici
Sez. Bari, c/o Dip. Chimica
70126 Bari, Italy
Chapter 7

Stefania Fasciano
CIRCC
70126 Bari, Italy
Chapter 8

Claudio Garino
Department of Chemistry
NIS Centre of Excellence and
INSTM Reference Center
University of Turin
10125 Torino, Italy
Chapter 5

Aharon Gedanken
Department of Chemistry
Bar-Ilan University
Ramat-Gan 52900, Israel
Chapter 1

Cinzia Giannini
Istituto di Cristallografia
Sede di Bari
70126 Bari, Italy
Chapter 6

Diego Gianolio
Diamond Light Source Ltd
Harwell Science and Innovation Campus
OX11 0DE Didcot
United Kingdom
Chapter 5

Roberto Gobetto
Department of Chemistry
NIS Centre of Excellence and
INSTM Reference Center
University of Turin
10125 Torino, Italy
Chapter 5

Elena Clara Groppo
Department of Chemistry
NIS Centre of Excellence and
INSTM Reference Center
University of Turin
10125 Torino, Italy
Chapter 5

Carlo Lamberti
Department of Chemistry
NIS Centre of Excellence and
INSTM Reference Center
University of Turin
10125 Torino, Italy
Chapter 5

Cosimino Malitesta
Laboratorio di Chimica Analitica
Di.S.Te.BA
Università del Salento
73100 Lecce, Italy
Chapter 3

Eleonora Margapoti
Laboratorio di Chimica Analitica
Di.S.Te.BA
Università del Salento
73100 Lecce, Italy
Chapter 3

Gema Martinez-Criado
European Synchrotron Radiation Facility (ESRF)
38043 Grenoble, France
Chapter 5

Lorenzo Mino
Department of Chemistry
NIS Centre of Excellence and
INSTM Reference Center
University of Turin
10125 Torino, Italy
Chapter 5

Anna Moliterni
Istituto di Cristallografia
Sede di Bari
70126 Bari, Italy
Chapter 4

Ilana Perelshtein
Department of Chemistry
Bar-Ilan University
Ramat-Gan 52900, Israel
Chapter 1

Nina Perkas
Department of Chemistry
Bar-Ilan University
Ramat-Gan 52900, Israel
Chapter 1

Andrea Piovano
Institut Laue-Langevin (ILL)
F-38042 Grenoble Cedex, France
Chapter 5

Rosanna Rizzi
Istituto di Cristallografia
Sede di Bari
70126 Bari, Italy
Chapter 4

Luca Salassa
CIC biomaGUNE
20009 Donostia -San Sebastián, Spain
Chapter 5

Teresa Sibillano
Istituto di Cristallografia
Sede di Bari
70126 Bari, Italy
Chapter 6

Dritan Siliqi
Istituto di Cristallografia
Sede di Bari
70126 Bari, Italy
Chapter 6

Federico Squassabia
STEPBIO Srl
40133 Bologna, Italy
and
Zinsser Analytic GmbH
Eschborner Landstrasse 135
Frankfurt, Germany
Chapter 2

Marinella Striccoli
CNR-IPCF Istituto per i Processi Chimici e Fisici
Sez. Bari, c/o Dip. Chimica
70126 Bari, Italy
Chapter 7

Marco Truccato
Department of Physics
NIS Centre of Excellence
University of Turin
10125 Torino, Italy
Chapter 5

Table of Contents

Angela Dibenedetto and Michele Aresta

Introduction: Nano- (and micro-)materials and human wellbeing

This book gathers the texts of the lectures delivered at the School on "Synthesis and characterization of novel nanosized inorganic materials" sponsored by EUCheMS-Inorganic Chemistry Division and held in Bari, Italy on June 17–22, 2012. The book mainly covers the aspects of synthesis and characterization of nanomaterials, with some reference to the characterization of micro-materials. In particular, several solid-state techniques that may discover the surface or 3D properties of the materials are discussed in detail.

Isolable particles of size < 1–< 50 nm, known since 1857 [M. Faraday, *Philos. Trans. R. Soc. London*, 147, 145–153], show unique properties when they are applied in several fields. Such non-usual behavior is in general correlated with the huge surface area they have and with the fact that a large number of active centers are exposed on the surface of the particle. The methodologies of synthesis of nanoparticles are based on two convergent approaches based on either the effective crushing of macro-particles or the controlled aggregation of atoms so as to obtain "clusters of a few atoms" which may also reach a size of less than 1 nm.

Nanoscience and nanotechnology are playing an ever increasing role in our lives as nanostructured materials with their unique mechanical, thermal, bio-medical and catalytic properties find application in many fields. Besides industrial uses as catalysts (synthesis of molecular or polymeric compounds, oxidation and hydrogenation reactions) or in fuel-cells, nano-materials are used in biomedical applications, textiles, electronics, automotive parts, food, among others. This is raising concerns about the safety of nanosized materials and the understanding of their potentially harmful or toxic effects is now an area of large debate and is attracting enormous scientific interest. In fact, nanoparticles because of their size can easily cross the cell membrane and disturb their correct growth.

In order to take full advantage of the enormous potential of nanomaterials' contribution to sustainable growth guaranteeing consumers wellbeing and safety, increasing interest is being put in the knowledge of the stability and reliability of nanoparticles. Therefore, techniques that may characterize nanoparticles and can detect them also in complex food- or bio-matrices are under study with ever greater interest.

The pervading use of nanomaterials has also raised concerns for the potential risks linked to the preparation phase for the exposure of workers.

Hopefully, so far, no alarming information has come from the research world as no serious toxic effects for humans have yet been discovered relevant to short-time exposure during the synthesis and use of nanomaterials. This safety aspect must be confirmed for long-time exposure.

Nanoparticles are now incorporated in more than 1300 commercial products [A. Maynard and E. Michelson, "The Nanotechnology Consumer Products Inventory", Washington, DC, Project on Emerging Technologies, 2011] and their use is foreseen to expand to an ever larger number of applications. The large investment in this area is now producing its effect with great benefit for human beings.

Aharon Gedanken, Nina Perkas, and Ilana Perelshtein

1 Coating antibacterial nanoparticles on textiles: Towards the future hospital in which all textiles will be antibacterial

1.1 Introduction: Application of nanotechnology for "smart" textiles

Nowadays, antibacterial properties have become one of the most essential characteristics for high quality textiles. With improvement of people's living standard, the requirements for hygienic and active lifestyle clothing include also nontoxicity, water resistance and long durability of the antibacterial function. The greatest challenge to researchers and manufacturers in the textile industry is development of effective ecofriendly technology for production of new multifunctional materials.

Worldwide industry reports estimate the wound care market exceeded $11.8 billion by 2009 and projected a yearly growth for all products (devices for wound closure such as sutures and staples, dressings, *etc.*) in excess of 7 %. European markets have accounted for about half of the spending [1]. Recent achievements in the field of antibacterial textiles were briefly described in a review by Gao [2]. According to his data, metal and metal salts are one of the major classes of antibacterial agents as well as quaternary ammonium compounds, triclosan, chitosan, chlorine-containing N-halamine compounds *etc.*, which are not considered as environmentally friendly substances.

Modern nanotechnology is an emerging interdisciplinary technology that has been booming in many areas during the last decade, including materials science, mechanics, electronics, and aerospace. Its profound societal impact has been considered as the huge momentum to usher in a second industrial revolution.

The impact of nanotechnology in the textile finishing area has brought up innovative finishes as well as new application techniques. Particular attention has been paid in making chemical finishing more controllable and more thorough. Ideally, discrete molecules or nanoparticles of finishes can be brought individually to designated sites on textile materials in a specific orientation and trajectory through thermodynamic, electrostatic or other technical approaches.

"Nano-functional" textiles are those designed and engineered on the nanoscale to create specific functions. These nano-functions are very diverse and can be used for:

- UV protection
- flame retardant
- moisture management
- antibacterial functions
- antistatic
- stain-resistant

Nowadays there are growing requirements for high-quality textiles with antibacterial properties for hygienic clothing, active wear, and wound healing. Last time control on microorganisms extends from hospital institutions to households. It is recognized that neither synthetic, nor natural fibers have resistance to bacteria and pathogenic fungi. An explosive growth is expected in wound care production. The wound care market of the US healthcare system was in excess of $7 billion in 2007. Consumer demand is causing significant growth in the production of the antibacterial textile.

Nanoparticles (NPs) such as metal oxides and ceramics are used in textile finishing altering surface properties and imparting textile functions. Nanosized particles have a larger surface that increases their area of contact and makes them more effective than larger size particles. These unique properties have found wide application in the textile industry, namely, in antibacterial treatment of textiles [3–5]. The market for textiles using nanotechnologies is predicted to climb dramatically from $13.6 billion in 2007 to $115 billion by 2012 [6].

Nanosilver is one of the most widely-used nanoparticles as an effective antibacterial agent in general textiles and in wound dressings [7–9]. The antibacterial properties of silver have been known and used for centuries [16]. A unique and available source of silver has long been mineral salts. A new way for delivery of silver into the bacterial killing medium is the formation of organic–inorganic nanocomposites combining the properties of textile substrate with antibacterial activity [10, 11]. To achieve the optimum antibacterial effect of nanocomposite fibers, a high concentration of silver ions must be available in the solution. Despite the small number of silver ions released from metallic silver nanocrystals, about 30 times less than that from silver complexes (*e.g.*, silver sulfadiazine), a more rapid microbe-killing curve has been observed with nanocrystals [12].

Different methods have been used for the deposition of silver NPs on fabrics. For example, a poly(ethylene terephthlate) fabric (meadox double velour) was coated with metallic silver using a patented ion-beam-assisted deposition process developed by the Spire Corporation (Bedford, MA). Antimicrobial fibers were produced by the implementation of nanoscaled silver particles into a solution of cellulose and N-methylmorpholine-N-oxide [13]. Other methods were constant pressure padding, impregnation in the colloid silver solution [14], immersion of the fabric in the silver precursor solution in ethanol or propanol following the boiling procedure for reduction of silver ions [15, 16], magnetron sputter technique [17] *etc.* Some of the methods are based on reactions in the liquid medium and require surfactants, reducing agents or templates for the synthesis of silver nanoparticles, resulting in the presence of toxic impurities in the final products. This method has some disadvantages with regard to the environment.

Metal nano-oxides such as ZnO, CuO and TiO_2 possess photocatalytic ability, UV absorption, and photo-oxidizing capacity against chemical and biological species. In the last decade intensive research involving metal nano-oxides was pushed forward focusing on the production of a textile with antibacterial, self-decontaminating and

UV-blocking functions [18, 19]. Nylon fibers filled with ZnO NPs can provide a UV-shielding function and reduce static electricity of nylon fiber [20]. A composite fiber with TiO_2/MgO NPs can provide a self-sterilizing function [21].

The nanostructured metals and inorganic oxides can be incorporated into the textile by various methods. Among the techniques are: high energy γ-radiation and thermal treatment assisted impregnation [22, 23]. In these works, cotton and cotton/polyester fabrics were immersed in antimicrobial formulation based on zinc oxide (ZnO), Impron MTP (binder), and Setamol WS (dispersing agent) and subjected to fixation by γ-radiation techniques. The effect of this treatment on the growth of bacteria (*Bacillus subtilis*) was estimated. On the basis of microbial detection, it was found that the ZnO formulation causes a net reduction in the bacterial cells which amounts to 78 % and 62 % in the case of treated cotton and cotton/polyester fabrics, respectively. However, it was found that the treatment with ZnO formulation caused a reduction in the thermal stability of the fabrics as indicated by thermogravimetric analysis.

One of the widely-used techniques for coating the textile substrates is the combination of sol-gel synthetic procedure with the "pad-dry-cure" method [23, 24]. The synthesis process usually involves two main steps. For instance, the hexagonally ordered ZnO nanorod arrays might be grown onto fiber substrates in the same way as zerogel ZnO [26]. The growing seeds were formed by coating ZnO nanosol using dip-coating, dip pad-curing or spraying methods by natural solvent evaporation. In order to stabilize the precursor solution, triethylamine with the same molar ratio as zinc acetate was added to form a transparent homogeneous solution.

The TiO_2 and TiO_2/SiO_2 nanocomposites prepared by low temperature sol-gel synthesis were coated onto cotton fabrics by a dip-pad–dry-cure process [27, 28]. The sol-gel immobilization and controlled release of various bioactive liquids from modified silica coatings were investigated in [29]. Deposition of nano-ZnO onto cotton fabric was performed by padding of the textile in the colloid formulation of zinc oxide-soluble starch nanocomposite to impart the material the antibacterial and UV-protection functions [30].

Some publications reported on deposition of "*in situ*" formed metal oxide NPs on the textile fabrics. A superhydrophobic ZnO nanorod array film on cotton substrate was fabricated *via* a wet chemical route and subsequent modification with a layer of *n*-dodecyltrimethoxysilane [31]. ZnO nanoparticles were grown in situ on SiO_2-coated cotton fabric through the hydrothermal method. After water treatment at 100 °C or higher, the cotton fabric was covered with approximately 24 nm diameter needle-shaped ZnO nanorods, which had an excellent UV-blocking property [32]. These NPs were impregnated onto cotton fabrics by the "pad-dry-cure" method using acrylic binder.

Copper is one of a relatively small group of metallic elements which are essential to human health. These elements, along with amino and fatty acids and vitamins, are required for normal metabolic processes. Copper is considered safe to humans as demonstrated by the widespread and prolonged use of copper intrauterine devices

(IUDs). At the same time, there are not many publications on production and application of CuO-textile composite with the exception of Gabbay *et al.* [33, 34]. The copper-containing fibers of cotton and polyester prepared by these authors demonstrated significant antifungal and antimicrobial properties. They inserted the preliminary synthesized copper oxide powder into the polymer fibers during the master-batch stage, and impregnated the cotton by a multi-phase soaking procedure including treatment in formaldehyde.

To summarize, most of the methods employed for the deposition of nanostructured materials on the textile are based on the multistage procedure including the preliminary synthesis of NPs and application of some templating agents for their anchoring to the substrates. This approach is rather complicated and can result in the release of some toxic compounds into the wastes. Therefore, the current research is focused on the fabrication of *in situ* coated fabrics *via* ultrasound irradiation. The sonochemical method prevents the use of toxic binders and makes the coating procedure shorter, effective, and environmentally friendly.

The next part of the chapter covers the background of sonochemistry, the theory and the possible applications.

1.2 Sonochemical method for the synthesis of nanostructured materials and their adherence to solid substrates

Sonochemistry has been proven as an efficient method for synthesis of various kinds of nanoparticles [35, 36]. In this chapter, we will describe the unique properties that make ultrasound irradiation an excellent technique for the adherence of nanoparticles to a large variety of substrates. Ultrasonic chemistry is a research field where waves in the frequency range of 20 kHz–1 MHz are the driving force for chemical reactions. The reaction is dependent on the acoustic cavitation, *i.e.*, the formation, growth, and implosive collapse of the bubbles in the solution. Extreme conditions (temperature > 5000 K, pressure > 1000 atm and cooling rates $> 10^9$ K/sec) are developed when the bubble collapses, resulting in chemical reactions [37].

A theoretical explanation as to how 20 kHz ultrasonic radiation can break chemical bonds is given in several works [38, 39]. The question arising is how such a bubble can be formed, considering the fact that the forces required to separate water molecules to a distance of two Van-der-Waals radii would require a power of 10^5 W/cm. On the other hand, it is well known that in a sonication bath with a power of 0.3 W/cm, water is readily converted into hydrogen peroxide. The explanation of this phenomenon is based on the existence of unseen particles or gas bubbles that decrease the intermolecular forces, enabling the creation of the void. The experimental evidence for the importance of unseen particles in sonochemistry is that when the solution undergoes ultrafiltration before the application of ultrasonic power, there is no chemical reaction and chemical bonds are not ruptured. The second stage is the growth of the bubble, which occurs through the diffusion of the solvent and/or solute vapors into the volume of the bubble. The third stage is the collapse of the bubble, which occurs when the size of the bubble reaches its maximum value. From here on we will adopt the hot spot mechanism, one of the theories that explain why upon the collapse of a bubble, chemical bonds are broken. This theory claims that very high temperatures (5,000–25,000 K) [40] are obtained upon the collapse of the bubble. Since this collapse occurs in less than a nanosecond, very high cooling rates in excess of 10^{11} K/sec are obtained. These extreme conditions develop when the bubble's collapse causes the chemical reactions to occur. The high cooling rate prevents the crystallization of the products. This is the reason why amorphous nanoparticles are formed when volatile precursors are used and the gas phase reaction is predominant. However, from this explanation the reason for the formation of nanostructured material is not clear. Our explanation for the creation of nanoproducts is that the fast kinetics does not permit the growth of the nuclei, and in each collapsing bubble a few nucleation centers are formed whose growth is limited by the short collapse. If the precursor is a non-volatile compound, the reaction occurs in a liquid phase in a 200 nm ring surrounding the collapsing bubble [39]. The products are sometimes nano-amorphous particles and in other cases nanocrystalline, depending on the temperature in the ring region where the reaction takes place. In fact, when the sonochemical reactions were used for the synthesis of inorganic products, nanomaterials were obtained.

Over the last 15 years we have reported on the ultrasound-assisted synthesis of about 100 nanomaterials and on the deposition/insertion of nanoparticles on/into ceramic and polymer bodies (see previous reviews) [41]. Other review articles on similar topics have also been published [42, 43].

In the current review we report on development of a sonochemical technique for the doping of produced NPs into the textile materials. We benefit from a well-known property of the acoustic bubbles, *i.e.*, when they collapse near a solid surface, microjets and shock waves are among the after-effects of the bubbles' collapse. These microjets throw newly-formed NPs onto textile at a very high speed, causing them to be embedded in the fabric.

Typically, the doping procedure is as follows. A textile substrate is introduced into the sonication cell containing the precursor solution. The ultrasonic irradiation passes through the sonication slurry under an inert or oxidizing atmosphere for a specified time leading to fabrication of NPs. This synthetic route is a single-step, effective procedure. The excellent adherence of NPs to the substrate is reflected, for example, in the lack of leaching of NPs from the substrate surfaces after many washing cycles. Another way of embedding NPs into the solids is the sonochemical irradiation of preliminary synthesized or commercial NPs in presence of corresponding substrates. This technique is called the "throwing stones" mechanism, because the ultrasonic waves provide throwing of the already existing NPs onto the fabric. In both cases the NPs strongly adhere to the substrate and the coating/incorporation is stable. Further on, the advantages of sonochemistry as a one-step, environmentally-friendly method for deposition of NPs on different kinds of textiles such as cotton, wool, nylon, polyester *etc.* with no use of any binding agents will be demonstrated.

1.3 Ultrasound assisted deposition of metal nano-oxides on textiles and their antibacterial properties

1.3.1 Synthesis and deposition of CuO nanoparticles

The sonochemical approach has been applied for deposition of CuO NPs on the textile fabrics [44]. The process takes place as follows: the copper ions react with a solution of ammonia to form a deep blue solution containing $[Cu(NH_3)_4]^{+2}$ complex ions (Eq. (1.1)). This complex is hydrolyzed and crystalline CuO NPs are obtained (Eqs. (1.2), (1.3)). The process takes place under sonochemical irradiation. The CuO NPs produced by these reactions are thrown at the surface of the fabric by sonochemical microjets and deposited on the surface of the substrate.

$$Cu^{2+}_{(aq)} + 4NH_3 \cdot H_2O_{(aq)} \longrightarrow [Cu(NH_3)_4]^{2+}_{(aq)} + 4H_2O \tag{1.1}$$

$$[Cu(NH_3)_4]^{2+}_{(aq)}) + 2OH^-_{(aq)} + 4H_2O \longrightarrow Cu(OH)_{2(s)} + 4NH_3 \cdot H_2O_{(aq)} \tag{1.2}$$

$$Cu(OH)_{2(s)} \xrightarrow{\Delta} CuO_{(s)} + H_2O \tag{1.3}$$

Morphology of the fibers' surface before and after the deposition of copper oxide was studied by X-ray diffraction (XRD) and high resolution scanning electron microscopy (HR SEM). The XRD revealed the presence of CuO nanocrystals in monoclinic structure. The difference in morphology between pristine and coated cotton fabric is clearly demonstrated in Figure 1.1. The insert image in Figure 1.1b at higher magnification shows uniform coating with primary NPs in a very low nanometric range ($\sim$10–20 nm).

While Cu^{+2} is considered an environmentally-safe ion, a much more important and serious issue is the leaching of CuO NPs. Diffusion light scattering (DLS) and transmission electron microscopy (TEM) studies of washing solution after treatment of the CuO-coated fabrics in 0.9 wt% NaCl did not reveal the presence of any nanoparticles. This means that the sonochemically-deposited CuO NPs are strongly anchored to the textile substrate, probably due to local melting of the fibers at the contact sites. Similar results were obtained for coating various types of textiles, such as nylon, polyester, and composite types of textiles with CuO NPs.

The antimicrobial activity of the cotton bandages coated with CuO *via* ultrasonic irradiation was tested against the *E. coli* and the *S. aureus*. Detailed investigations showed that after 1 h the growth of both strains was completely inhibited. One of the factors influencing the antibacterial activity of the developed coating is release of the active phase, namely copper ions or/and copper oxide NPs, into surrounding medium. Examination of leaching of the copper ions indicated that only a very low amount

Fig. 1.1: HR SEM images of cotton fibers: (a) pristine cotton; (b) cotton coated with CuO NPs (magnification $\times$ 20,000, inset images – magnification $\times$ 100,000).

(namely $\sim$1.3%) of the deposited copper was removed by washing with a 0.9 wt% NaCl solution that corresponds to a concentration of Cu^{2+} 0.15 ppm. Slight solubility of copper oxide can be explained by very low K_{sp} of CuO, which dictates that the very small concentration of Cu^{+2} appears in the washing solution. In order to examine the influence of copper ions on antibacterial effect, a control antibacterial test was performed using a supernatant with the same concentration of Cu^{2+} instead of the coated cotton. After incubation for 24 h at 37 °C, no reduction in *E. coli* after 2 h was observed (Fig. 1.2).

This result indicates that the Cu^{2+} ions have no influence on the antibacterial activity. Thus, the antibacterial effect can be attributed to the copper oxide NPs. Although CuO nanoparticles were not found in the solution, they can generate some active species that are responsible for damaging of the bacteria's cells. These active species were detected in ESR studies conducted with and without the bacteria present in the ESR tube.

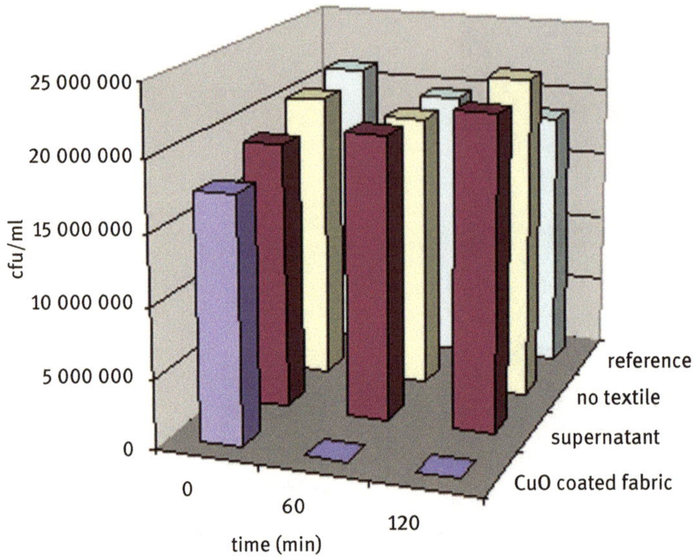

Fig. 1.2: Antibacterial test of CuO-coated cotton against *E. coli*.

1.3.2 Finishing of textiles with crystalline TiO_2 nanoparticles via a one-step process

Titanium oxide (TiO_2) NPs in their two crystalline forms, anatase and rutile, were synthesized from an aqueous solution of tetraisopropyltitanate (TPT) and titanium tetrachloride (TTC), respectively. Synthesis and deposition on the surface of cotton fabrics was done by using ultrasonic irradiation, in one-step reaction [45]. One of the aims of this research was to reach a minimal effective concentration of the deposited metal oxide NPs on the fabrics still demonstrating antimicrobial activity. The results indicated that the amount of coated titanium oxide depends on the concentration of TPT/TTC in the solution and the sonication time. Decrease in the reaction time from 3 to 1 h led to a reduction in the content of TiO_2 deposited on the fabric, by approximately a factor of two. The decrease in the initial concentration of the TPT/TTC precursor also resulted in less titanium oxide in the nanocomposites. Varying the initial precursor concentration and reaction time one can control the content of TiO_2 in the composite.

The morphology of the fibers coated using TTP as a precursor and the structure of the deposited TiO_2 NPs (anatase form) were investigated by TEM with conventional selected area diffraction (SAED) technique, and STEM mode for elemental analysis and mapping (Fig. 1.3). A few individual TiO_2 crystalline NPs are clearly visible in the bright field (Fig. 1.3a), and in the dark field (Fig. 1.3b) images. The particle size is in the low nanometric range ($\sim$ 10–15 nm). The dark field image (Fig. 1.3b) was taken in the (101) direction, corresponding to the main reflection line of the body-centered tetragonal structure of anatase. The identification of the NPs was based on the analysis of the SAED pattern (c). The electron diffraction pattern was indexed in terms of a tetragonal unit cell of anatase TiO_2 (a = 3.78 Å, c = 9.5 Å).

Identification of the NPs was also supported by elemental analysis of the coated fiber. Energy dispersive X-rays (point EDS) and mapping were performed in STEM mode (Fig. 1.3d). Elemental mapping techniques for the analysis of distribution of TiO_2 NPs over a coated fiber were conducted in STEM mode. Figure 1.3e depicts Ti mapping showing Ti-containing particles covering the fiber. The Ti-mapping technique confirmed results presented in the TEM micrograph (Fig. 1.3a) and demonstrated that the Ti-containing NPs covered uniformly all fiber.

EDS spectrum (Fig. 1.3f) measured from the coated fiber with a 35 nm electron probe shows presence of Ti, O, and C elements in the analyzed area. The Cu peak originates from the specimen grid.

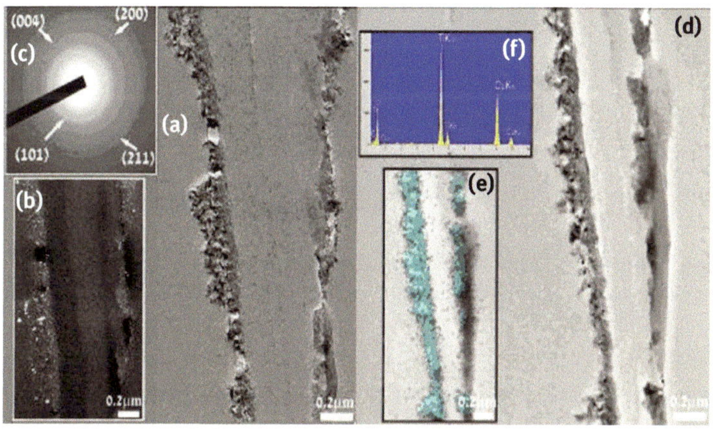

Fig. 1.3: HR TEM images of the TiO_2-fabric nanocomposite: (a) bright field; (b) dark field; (c) diffraction pattern; (d, e) image in the STEM mode (Ti is shown as blue dots); (f) EDS spectrum.

Morphology of the TiO_2-coated yarn was studied by HR SEM (Fig. 1.4; sample 6, rutile form). Figure 1.4a demonstrates a smooth texture of pristine cotton bandage. Following the sonication process, fibers were homogeneously coated with NPs (Fig. 1.4b). Figure 1.4c presents a higher magnification of the coated sample, which allows obtaining the particle size distribution. Distribution of the particles is quite narrow, and primary particles are in a very low nanometric range (~ 10–15 nm).

One of important issues examined during the above-mentioned research is the leaching of active phase (NPs) from the coated fabrics. The bandages coated with 2.3 % of TiO_2 (wt%) were treated with an aqueous solution of sodium chloride. A piece

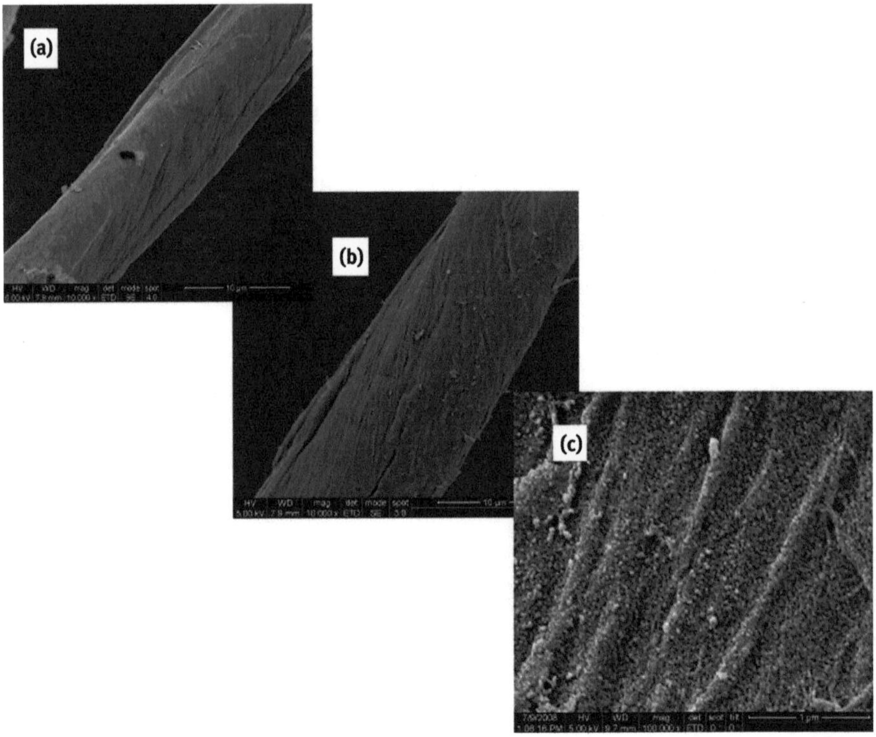

Fig. 1.4: HR SEM images of the fibers: (a) before coating at a magnification of 10 K; (b) after TiO_2 coating at a magnification of 10 K. Image (c) was taken at magnification of 100 K.

(0.08 g) of the coated bandage was kept in 0.9 % NaCl solution overnight at 37 °C and content of the Ti^{+4} in liquid was analyzed by ICP. Concentration of Ti^{+4} ions found in the leaching solution was 0.13 ppm. Thus, only $\sim$ 0.6 % of the active phase was removed by washing in NaCl solution. This result indicated that there is almost no solubility of the active phase into the washing solution. In addition to ICP measurements, the leaching solution was probed for the presence of NPs by DLS and TEM. According to the results, no presence of any NPs in the leaching solution was detected. The combination of the above-mentioned techniques proved that the sonochemically-deposited TiO_2 NPs are strongly anchored to textile substrate.

Photoactivity of the TiO_2 deposited on the cotton textile was estimated by the discoloration of the MB solutions exposed to UV irradiation in presence of the immersed titania-coated fabrics. Experiments on methylene blue (MB) discoloration under UV light demonstrated that the structure of titania NPs deposited on the cotton fabric is a determining factor for its photocatalytic activity. The anatase-coated cotton was more active in MB photo degradation than the rutile-coated sample (Fig. 1.5). This result is not unexpected, however, the influence of structure and morphology of the rutile NPs

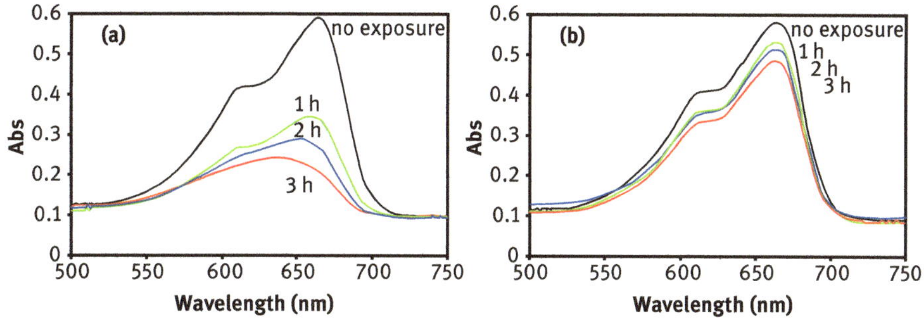

Fig. 1.5: Absorption spectra of an MB solution in presence of the immersed TiO_2-coated cotton fabrics during UV irradiation: (a) anatase structure of titania; (b) rutile structure of titania.

on its photocatalytic properties was also reported [46, 47]. In our study, we observed that the rutile NPs are of very high dispersion (10–15 nm), well distributed on the fibers (Fig. 1.4c). Therefore, the question raised is whether the TiO_2 (rutile)-cotton composite possessed at least some photocatalytic activity.

It was observed that with the samples of TiO_2 (rutile) cotton, the discoloration of MB took place at a much slower rate than with a TiO_2 (anatase) cotton composite, but faster than without any catalyst on the fabric in a control experiment (Fig. 1.6).

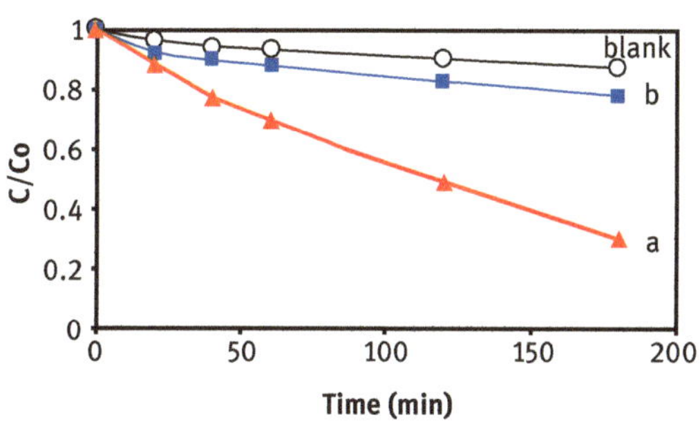

Fig. 1.6: Photocatalytic degradation of an MB solution in presence of TiO_2-coated cotton upon UV irradiation; (a) anatase structure of titania; (b) rutile structure of titania. Blank-control experiment of untreated cotton fiber.

Based on the obtained results, the TiO_2 (anatase) cotton composite prepared by an ultrasound-assisted technique can find wide application in the production of the self-cleaning fabric.

There is some controversy in the literature regarding the effective inactivation of microorganisms by TiO_2, primarily due to irradiance (UV/vis) of different experimental conditions, length of exposure, photocatalysts presented in suspension or in powder, range of concentration, and different TiO_2 photocatalysts and microorganisms employed, although some effect is generally acknowledged [48]. The bactericidal effect of titania was reported on food-pathogenic bacteria such as *Salmonella choleraesuis*, *Vibrio parahaemolyticus*, and *Listeria monocytogenes* [49], as well as *Pseudomonas aeruginosa* [50]. Forming a well-adhering bactericidal surface of TiO_2 on organic cellulose fibers was studied by Daoud [27]. Kangwansupamonkon *et al.* reported on the antibacterial performance of apatite-coated TiO_2 which was fixed on cotton textiles by a dip-coating technique [51]. Their study indicated that photocatalytic activity of an apatite-coated TiO_2 suspension can help in microbial decomposition in textile applications. Its effectiveness was clearly confirmed against *S. aureus*, *E. coli* ATCC 25922, MRSA DMST 20627, and *M. luteus* strains of bacteria. Also, Wu *et al.* reported on self-cleaning fabrics which had been prepared by depositing and grafting TiO_2 NPs *via* nonaqueous sol process at low temperature [52].

We determined antimicrobial activity of TiO_2 coated on cotton fabric using prokaryote cells, Gram-positive bacterium *S. aureus* and Gram-negative bacterium *E. coli*, as well as eukaryote cells, *Candida albicans*. In another test, the samples were illuminated with a visible light source at 450 nm (100 mW/cm^2 light emitting diode). It should be noted that incubating pristine bandage with examined strains in saline solution did not show any effect on their viability, even after 4 h of incubation. As depicted in Table 1.1, *S. aureus* bacterium was more susceptible than *E. coli* for treatment of the TiO_2-coated fabric for both rutile and anatase. Regarding *S. aureus*, almost a 100 % reduction in viability was reached after 3 h in the presence of light. Generally, the best antibacterial activity could be obtained under illumination. Furthermore, although anatase was found to be more effective than rutile against bacteria and candida, it was surprising that illumination enhanced the antimicrobial activity of rutile in a more effective manner than that of the anatase. Notably, both forms of titania, rutile and anatase, implicated lower antimicrobial activity in the case of *E. coli*, and *Candida albicans* in comparison to *S. aureus* (Tab. 1.1). This result is contradictory to our results obtained for ZnO and CuO where a better killing of *E. coli* was detected.

Concerning the illumination effect on antimicrobial activity of the titania-cotton composite, it has already been established that upon irradiation of a TiO_2 surface with photons, an electron is promoted from the valence band to the conduction band, thus forming an electron-hole pair. The photo-generated holes and electrons can easily react with the water molecules attached to TiO_2 surfaces in the presence of oxygen to form hydroxyl radicals and other ROS such as superoxide ions. It is still a subject of investigation as to which of these ROS are directly involved in the damage to bacteria cells. Moreover, until recently it remained unclear what conditions of irradiated light necessitate the activation of ROS generation.

Recently, we investigated the influence of visible light on the inducement of ROS in a titanium dioxide (TiO_2) water suspension using electron spin resonance (ESR) coupled with the spin-trapping probe technique. We detected high levels of hydroxyl (OH) and superoxide anion (O_2^-) radicals in water suspensions of TiO_2 rutile and anatase

Tab. 1.1: Antimicrobial activity test of the coated cottons against *E. coli*, *S. aureus*, and *Candida albicans*. Viable microorganism cells were monitored by counting the number of colony-forming units (CFU); $N/N0$ is the survival fraction.

	Duration of treatment					
	1 h			3 h		
Sample	**CFU ml^{-1}**	**N/N_0**	**% reduction in viability**	**CFU ml^{-1}**	**N/N_0**	**% reduction in viability**
			E. coli			
Clean fabric	7.2×10^7	0.99	1.0	7.2×10^7	0.99	1.0
Rutile	6.9×10^7	0.96	4.2	6.7×10^7	0.93	6.9
Rutile + light	6.3×10^7	0.87	12.6	4.9×10^7	0.68	31.9
Anatase	6.6×10^7	0.92	8.3	6.3×10^7	0.75	25
Anatase + light	6.3×10^7	0.87	12.6	5.1×10^7	0.71	29.2
			S. aureus			
Clean fabric	6.7×10^7	0.99	1.3	6.7×10^7	0.99	1.3
Rutile	2.7×10^7	0.40	59.6	1.9×10^7	0.28	72.4
Rutile + light	1.9×10^7	0.28	71.6	4.6×10^5	6.87×10^{-3}	99.3
Anatase	1.2×10^7	0.18	82.3	3.8×10^5	5.67×10^{-3}	94.4
Anatase + light	9.3×10^6	0.14	86.1	6.5×10^2	9.76×10^{-6}	99.9
			Candida albicans			
Clean fabric	8.7×10^8	0.99	0.8	8.7×10^8	0.99	0.8
Rutile	8.2×10^8	0.95	5.3	5.2×10^8	0.60	40.3
Rutile + light	5.9×10^8	0.68	32.2	5.4×10^8	0.38	62.4
Anatase	7.8×10^8	0.9	9.9	3.5×10^8	0.41	59.5
Anatase + light	5.14×10^8	0.59	40.9	2.6×10^8	0.30	70.1

NPs, with the signal monitoring both OH and O_2^- radicals being higher in the anatase phase without any light illumination conditions. However, illumination with visible (non-UV) light enhanced largely O_2^- formation in the rutile phase. Singlet oxygen was not detected, neither in the water suspension of TiO_2-rutile nor in the anatase-structured NPs, but irradiation of the rutile phase with visible light revealed a signal which was attributed to singlet oxygen formation. The blue part of the visible spectrum (400–500 nm) was found to be responsible for the light-induced ROS in TiO_2 NP suspensions. Furthermore, Kispert and co-workers have shown that carotenoids act as photosensitizers for TiO_2 colloids, shifting the absorption band to the longer wavelengths and capable of generating the O_2^- and singlet oxygen on visible light (550 nm) irradiation [53]. Photosensitization of TiO_2 by carotenoids facilitates generation of the superoxide radical anion, O_2^-. These findings, together with the fact that *S. aureus* strains are equipped with carotenoid pigment [54], may explain the increased susceptibility of these strains towards titania treatment. A detailed paper concerning the issue of ROS production by TiO_2 NP water suspensions is presently in preparation. Our explanation for the enhanced antimicrobial killing resulting from the illumination of the sample by visible light is due to the absorption of impurity (defects) states below the bottom of the conductive band of titania. As mentioned above, these defect states are responsible for the formation of ROS. Populating these excited states is done by visible light.

1.3.3 Synthesis and deposition of ZnO

It is well known that the antibacterial activity of ZnO depends on the particle size: decreasing the particle size leads to an increase in antibacterial activity [55]. Our aim in work on coating cotton bandage was to get the most homogeneous coating of highly dispersed ZnO NPs and to reach a minimal effective ZnO concentration, which will still demonstrate antibacterial activity. We have developed a simple and effective method for immobilizing ZnO NPs into fabrics *via* ultrasound irradiation [56] and its simultaneous deposition on fabrics in a one-step reaction.

It was found that yield of product and particle size are strongly dependent on the rate of inter-particle collisions and on the concentration of reagents in the sonochemical synthesis. That is why the experimental parameters such as time and concentration of precursor were selected as important factors for optimization of the sonochemical reaction.

XRD demonstrated that ZnO NPs on the coated bandage are crystalline, and observed that diffraction patterns matched the hexagonal phase of ZnO structure. No peaks characteristic of any impurities were detected. The particle size estimated by the Debye–Scherrer equation is 30 nm. Morphology of the coated bandage before and after deposition of ZnO NPs studied by HR SEM is presented in Figure 1.7. In contrast to the smooth texture of pristine cotton bandage, after sonication the fibers of bandage are homogeneously coated with NPs (Fig. 1.7b). The inset image Figure 1.7b was taken at a higher magnification in order to obtain the particles' size distribution. Distribution of the particles is quite narrow, and primary particles are in a very low nanometric range ($\sim$ 30 nm) that matches well with the XRD results. The selected-area HR SEM image studied with elemental dot-mapping technique is shown in Figure 1.7c. The distribution of zinc and oxygen in the mapped area verified a homogeneous coating of the fibers with ZnO NPs.

We considered the following mechanism of ZnO formation:

$$Zn^{2+}_{(aq)} + 4NH_3 \cdot H_2O_{(aq)} \longrightarrow [Zn(NH_3)_4]^{2+}_{(aq)} + 4H_2O \tag{1.4}$$

$$[Zn(NH_3)_4]^{2+}_{(aq)} + 2OH^-_{(aq)} + 3H_2O \longrightarrow ZnO_{(s)} + 4NH_3 \cdot H_2O_{(aq)} \tag{1.5}$$

Ammonia works as a catalyst for the hydrolysis process, and formation of zinc oxide takes place through the ammonium complex $[Zn(NH_3)_4]^{2+}$. The ZnO NPs produced by this reaction are thrown at the surface of bandages by the sonochemical microjets resulting from collapse of the sonochemical bubbles.

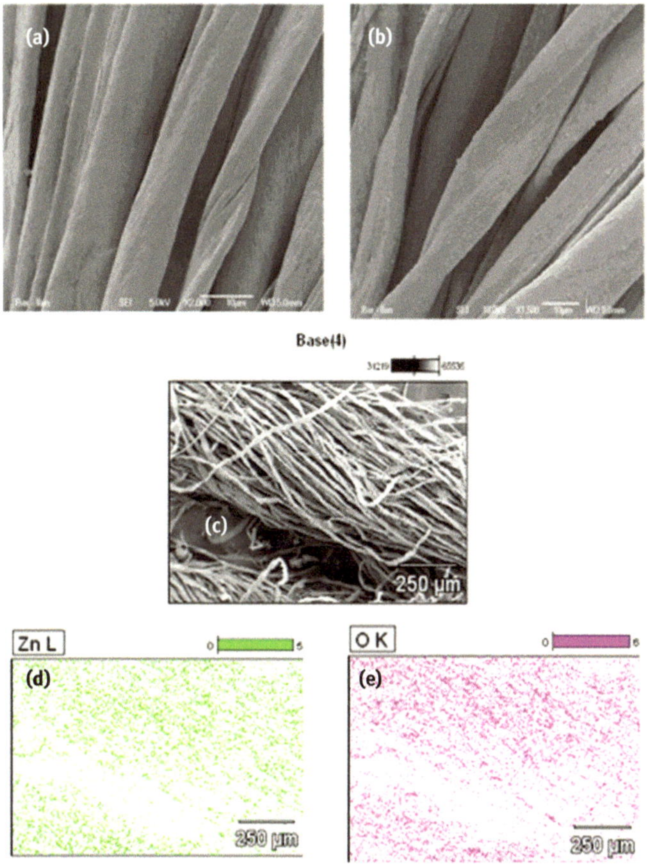

Fig. 1.7: HR SEM images of (a) pristine bandages (magnification × 2 K); (b) bandage coated with ZnO NPs (magnification × 1,500; the inset shows a magnified image (× 50 K) of the ZnO NPs on the fabric); (c) selected image for X-ray dot mapping; (d) X-ray dot mapping for zinc; (e) X-ray dot mapping for oxygen.

Sonochemical irradiation of a liquid causes two primary effects, namely, cavitation (bubble formation, growth, and collapse) and heating. When the microscopic cavitation bubbles collapse near the surface of the solid substrate, they generate powerful shock waves and microjets that cause effective stirring/mixing of the adjusted layer of liquid. The after-effects of cavitation are several hundred times greater in heterogeneous systems than in homogeneous systems. In our case, ultrasonic waves promote fast migration of newly formed zinc oxide NPs to the fabric's surface. This fact might cause a local melting of fibers at the contact sites, which may be the reason why the particles strongly adhere to the fabric. To further support the coating mechanism, FTIR spectra of pristine cotton bandage and ZnO-coated bandage have been recorded. Both spectra show the characteristic bands of cellulose. The recorded spectrum of ZnO-coated bandages revealed an additional sharp single band at 464 cm^{-1}, which is attributed to a Zn-O vibrational band. The coating assisted by ultrasound irradiation is a physical phenomenon that occurs regardless of the surface properties, especially when there are chemical interacting groups on the surface.

In this aspect, the question rises of if the sonication does not damage the fabric substrate. Thus, the tensile mechanical properties of a cotton impregnated fabric were studied on a universal testing machine, Zwick 1445. The tensile force for the zinc oxide-coated sample was observed to be ~11 % less than that of the pristine. The observed changes in mechanical behavior of the yarn are in a range that is acceptable for standard cotton fabrics. According to this result, we conclude that the sonochemical treatment of bandage does not cause any significant change in the structure of the yarns (Fig. 1.8).

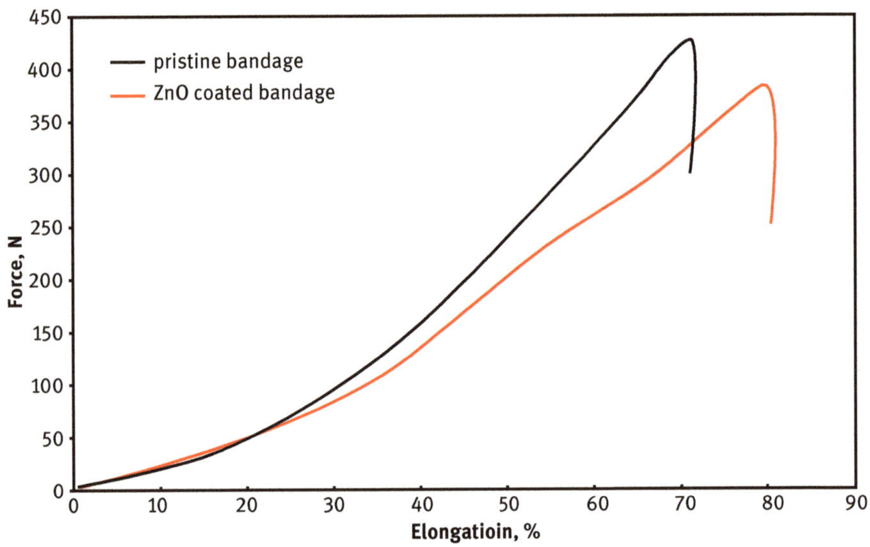

Fig. 1.8: Mechanical properties of cotton bandage before and after deposition of ZnO.

One of the factors influencing the commercial exploitation of antibacterial fabrics is the release of NPs into the surrounding environment. We attempted to find ZnO NPs in the washing solution. We carried out leaching experiments in 0.9 % NaCl (pH ~ 7) solution for 96 h, checking the presence of nanoparticles/ions in the solution after each 24 h. Figure 1.9 presents the change of percentage of ZnO remaining on fabric during the time. After 96 h 30 % of the initial concentration of Zn^{+2} was found in the washing solution. The reason for the presence of Zn^{+2} in the washing solution can be explained by the constant of solubility (K_{sp}) of the material that was estimated as $\sim 10^{-10}$.

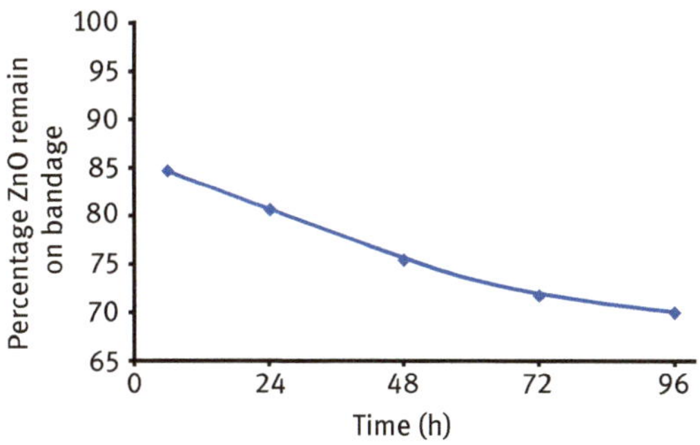

Fig. 1.9: Change of ZnO percentage remaining on bandages.

DLS and TEM studies did not reveal the presence of any nanoparticle in the washing solution after 96 h. This means that sonochemically deposited ZnO nanoparticles are strongly anchored to the textile substrate.

The ZnO-coated bandages were taken one step further – for sterilization. Eight samples coated with ZnO NPs in two concentrations, 0.8 % and 1.65 % (%wt), were tested by 4 different sterilization techniques: (1) gamma, (2) damp heat at 134 °C, (3) steam at 121 °C, (4) ethylene oxide (EO). After sterilization the samples were characterized for the quantity of ZnO coating, morphology of the NPs, and for their antibacterial activity. Table 1.2 summarizes the quantity results before and after sterilization. It can be seen that there is no significant change in the amount of coating after sterilization.

Tab. 1.2: Content of ZnO before/after sterilization.

Sample number	Average content of ZnO before sterilization	Average content of ZnO after sterilization
1-1	0.80	0.81
1-2		
1-3		
1-4		
2-1	1.65	1.60
2-2		
2-3		
2-4		

Morphology of the coating after sterilization was tested by SEM (Fig. 1.10). All 8 samples were tested, and Figure 1.10 presents only 2 of the obtained results a) sterilized by EO, and b) sterilized by steam at 121 °C. No significant difference such as aggregation of NPs was observed after exposure of the ZnO-coated bandages to different types of

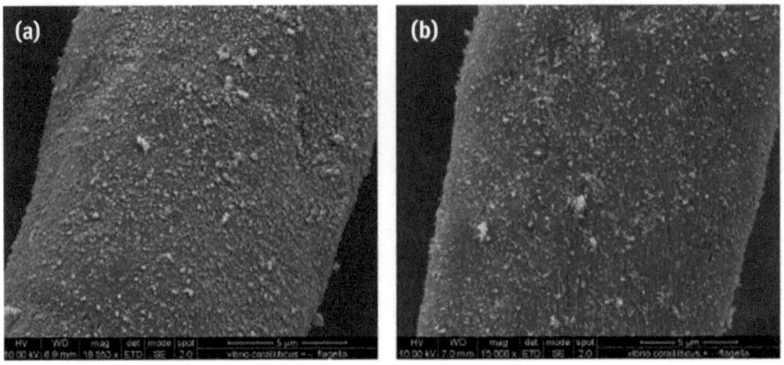

Fig. 1.10: SEM image of (a) cotton bandage sterilized by EO; (b) cotton bandage sterilized by steam at 121 °C.

sterilization. Obtained results indicated that the above-mentioned sterilization techniques did not influence quantity and morphology of the sonochemical coating. The NPs are strongly adhered to the fabric surface and remain stable after exposure to extreme conditions.

The mechanism of the antibacterial activity of the metal nano-oxides is poorly understood, and is still controversial. Suggested mechanisms in the literature include the role of the reactive oxygen species (ROS) generated on nano-oxides, ion release, membrane dysfunction, and nanoparticle internalization [57–59]. The generation of ROS is derived from the highly reactive nature of defect sites (such as oxygen vacancies) on a wet metal oxide surface.

In the work by Sawai *et al.* [60, 61], it was demonstrated that ROS concentration increased with the ZnO content of the slurries. Following the same paradigm, Applerot *et al.* [62], in an innovative study using electron spin resonance (ESR) coupled with the spin-trapping probe technique, monitored ROS, namely, hydroxyl radicals produced in water suspensions of ZnO NPs. These findings showed that the amount of hydroxyl radicals was closely related to the size of ZnO particles, with smaller sizes having greater amounts of OH on the basis of equivalent ZnO mass content. These results were correlated with an increase in the antibacterial effect of small NPs. Thus, the small size and large specific surface area endow them with high chemical reactivity and intrinsic toxicity. Interestingly, combining Gram-negative bacterium *E. coli* and ZnO NP suspensions immediately enhanced the generation of OH for up to an average of 142 % (Fig. 1.11).

Antibacterial activity of the sonochemically-coated cotton fabrics containing 0.75 wt% ZnO against the Gram-negative bacterium *E. coli* and the Gram-positive bacterium *S. aureus* is shown in Table 1.3. Treatment for 1 h with the coated cotton leads to complete inhibition of *E. coli* growth. Regarding *S. aureus*, 100 % reduction in viability was reached after 3 h, while after 1 h of treatment a reduction of 60 % could be seen.

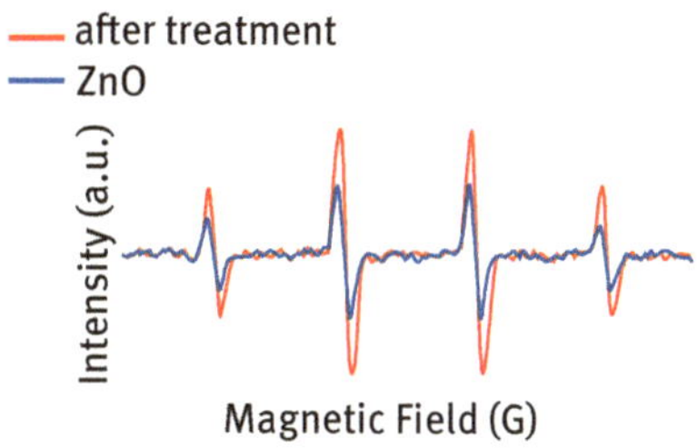

Fig. 1.11: ESR spectra demonstrating changes in hydroxyl radical concentrations upon antibacterial treatment of *E. coli* with a water suspension of ZnO.

Tab. 1.3: Antibacterial activity test of ZnO-coated cotton.*

	Duration of treatment					
	1 h			3 h		
Sample	**CFU ml^{-1}**	**N/N$_0$**	**% reduction in viability**	**CFU ml^{-1}**	**N/N$_0$**	**% reduction in viability**
E. coli						
Pristine fabric	$1.02 \cdot 10^{-7}$	0.98	0.98	$1.34 \cdot 10^{-7}$	1.28	−28.23
No fabric	$1.17 \cdot 10^{-7}$	1.14	−28.57	$1.23 \cdot 10^{-7}$	1.35	−35.16
0.75 wt%ZnO on fabric	$1.71 \cdot 10^{-7}$	$1.58 \cdot 10^{-3}$	99.84	0	$0.9 \cdot 10^{-8}$	100
	Duration of treatment					
	1 h			**2 h**		
Sample	**CFU ml^{-1}**	**N/N$_0$**	**% reduction in viability**	**CFU ml^{-1}**	**N/N$_0$**	**% reduction in viability**
S. aureus						
Pristine fabric	$0.7 \cdot 10^{-7}$	0.71	20.46	$0.99 \cdot 10^{-7}$	1.125	−12.5
No fabric	$0.98 \cdot 10^{-7}$	1.10	−10.11	$0.67 \cdot 10^{-7}$	0.75	24.72
0.75wt%ZnO on fabric	$3.9 \cdot 10^{-6}$	0.34	66.4	$7.6 \cdot 10^{-3}$	$6.55 \cdot 10^{-4}$	99.93

* The viable bacteria were monitored by counting the number of colony-forming units (CFU); N/No: survival fraction.

Zinc is an essential micronutrient for prokaryotic organisms. However, at super physiological levels, Zn^{2+} inhibits the growth of many bacteria [63]. According to our leaching experiment on the ZnO-embedded fabric with 0.9 wt% NaCl, the concentration of Zn^{2+} in solution corresponds to 36.7 μM/L [56]. Compared to the minimum inhibitory concentration reported in the literature of 4–8 mM/L [64], the amount of zinc ions released from fabrics in our work is lower at least by a factor of 2. Therefore, we assume that the Zn^{2+} have a minor influence on the antibacterial activity. The major components responsible for the bactericidal effect are the ZnO NPs. Although ZnO NPs were not found in the solution, they can generate some species of oxy-radicals as was reported earlier [65].

1.3.4 Enzymatic pretreatment as a means of enhancing antibacterial activity and stability of ZnO nanoparticles sonochemically coated on cotton fabrics

Controlled surface hydrolysis of cotton with cellulase enzymes, a process known as biopolishing, can modify the surface of fibers at low temperature [66] improving the texture, appearance, softness, luster and drape of the textile material [67–72]. Biopolishing is currently an industrial textile finishing process employed to remove protruding yarns from the surface of cotton fabrics [69]. The commercial cellulase formulations can contain mixtures of endoglucanases, exoglucanases and β-glucosidases exerting different effects on a fabric's properties. For instance, one component may decrease pilling, but reduce tear strength, while another component may have no effect on tear strength [68]. Cellulase pretreatment of cotton, to the best of our knowledge, has never been used as a method to improve the adherence and uniformity of NPs coatings.

The enzymatic treatment is expected to increase the surface area and porosity of fibers and to create more anchoring points, *e.g.*, end groups in cellulose, for NPs grafting onto the surface. Such surface modification of cotton fabrics may provide better adhesion and uniform distribution of the ZnO NPs, and thus ensure stability of the antibacterial effect during exploitation.

We produced ZnO NP coated textiles after their pretreatment with cellulase and studied the effect of the enzymatic pretreatment on size, distribution and antimicrobial activity of the sonochemically deposited nanoparticles [73]. The sonochemical deposition process was performed on a small pilot machine, which we had previously developed and used to coat various substrates in a roll-to-roll mode [74]. In order to maintain a constant concentration of the ZnO NP coating over the entire length of the textile specimen, concentration of the solution was maintained by adding the necessary amount of ZnO precursor to the sonication cell during the procedure. In this way NPs were formed continuously to provide a uniform coating over the complete cotton fabric.

The amount of ZnO deposited on the fabric was determined by ICP measurements. In Table 1.4 the amount of ZnO along the cotton fabric pretreated with 3 % cellulase o.w.f. is presented as a function of the distance from the starting point. The average concentration of ZnO along 10 m coated fabric was 0.7 ± 0.04 wt%.

The cotton fabrics not treated with cellulase were coated with ZnO using the same reaction conditions as for the enzyme pretreated textiles. The average amount of ZnO on the nonpretreated fabric was 0.75 ± 0.04 wt% as determined by ICP.

Tab. 1.4: Concentration of ZnO along the fabric (experimental error is ±6 %).

Position along the textile (m)	Concentration of ZnO (% wt)
1	0.72
2	0.70
5	0.68
10	0.66

The XRD method determined the presence of crystalline ZnO in hexagonal phase in the coated fabric. No peaks characteristic of any impurities were detected.

According to SEM measurements (Fig. 1.12), after the cellulase pretreatment the fiber's surface became more uniform (Fig. 1.12d), even, and hairless in comparison to the surface of the pristine fabric (Fig. 1.12a). This was due to the hydrolytic removal of protruding fibrils and primary walls of the fibers. The SEM micrographs also revealed a difference between deposition of ZnO NPs on the enzyme pretreated and nontreated fabric. A highly dense coating of ZnO NPs on the nontreated fibers was observed (Fig. 1.12b). Figure 1.12c depicts a fragment of the coated fabric from Figure 1.12b imaged under higher magnification. Most of the nanoparticles were in the size range of 60–70 nm. At the same time, the tendency of particles to form small

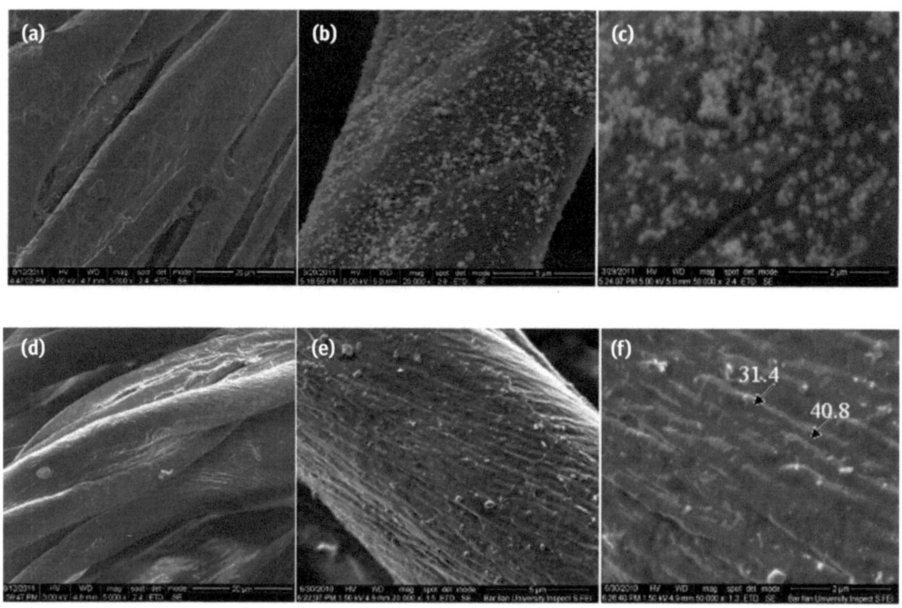

Fig. 1.12: SEM images of: (a) pristine cotton fibers (MAGX5K); (b) cotton fibers coated with ZnO NPs (MAGX20K); (c) a higher magnification of B (MAGX50K); (d) cotton fibers pretreated with 3 % cellulose o.w.f. (MAGX5K); (e) cotton fibers pretreated with 3 % cellulase o.w.f. and coated with ZnO NPs (MAGX20K); (f) a higher magnification of E (MAGX50K).

aggregates of around 100 nm was also observed. In contrast, the fabrics coated with ZnO NPs after preliminary treatment with 3 % cellulase o.w.f. (Fig. 1.12e) showed uniform nanoparticle dispersion along the fibers, and particle sizes in the lower nanometric range of 30–40 nm and even smaller particles were visible (Fig. 1.12f). Based on the above-mentioned results it can be concluded that the pretreatment of cotton fabrics with cellulase influenced the morphology of the sonochemical deposition of NPs. The enzymatic pretreatment caused a reduction of the particle size and prevented NP aggregation. It was previously demonstrated that the particle size of the NPs had a major effect on antibacterial properties and smaller NPs were more efficient biocides.

It is known that cellulase, endoglucanase (EG) will hydrolyze randomly the β-(1-4) glycoside bonds in the amorphous regions of cellulose, decreasing the degree of polymerization (DP) of the polymer and increasing the number of reducing chain ends in cellulose macromolecules. This will influence the reactivity of cellulose and can provide more anchoring points for NP attachment and growth. Copper number determination was used to estimate the quantity of reducing ends in cellulose, thus providing information about the chemical changes and new functional groups appearing on the fiber surface. Normally endoglucanases cause a significant increase of the copper number of cotton fabrics. Increasing the concentration of cellulase resulted in up to a 3-fold increase of the copper number for the samples treated with 3 % cellulase o.w.f. as compared to the untreated fabric (Fig. 1.13). Further increases in enzyme concentration (4–5 %) did not significantly change the copper number. According to this data, pretreatment of the fibers provided additional free hydroxyl groups, *i.e.*, reactive sites serving as nucleation centers for the NPs.

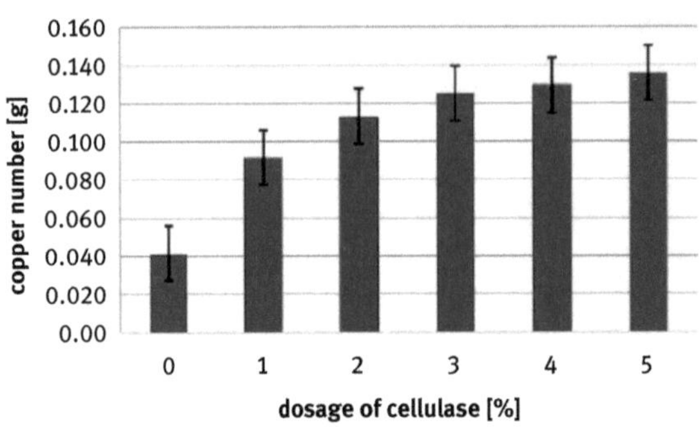

Fig. 1.13: Copper number (the number of grams of metallic copper obtained from the reduction of $CuSO_4$ by 100 g of cellulose fibers) of the samples treated with different concentrations of cellulase (% o.w.f.), the baseline corresponds to amount of copper number of control fabric (without cellulase treatment) and is equal to 0.42 g.

Although the hydrolytic enzymatic treatment may cause loss of mechanical strength of the fabrics, generally, a controlled enzymatic attack results in only moderate weight and strength losses. Concentration dependent weight loss on cotton fabrics was observed after cellulase treatment (Fig. 1.14). Pre-treatment with 3 % o.w.f. cellulase was selected for further sonochemical deposition of ZnO NPs because it gave rise to an acceptable weight loss in the fabrics. A weight loss of 1.5–2.5 % is not considered detrimental for the tensile strength of the fabric [67].

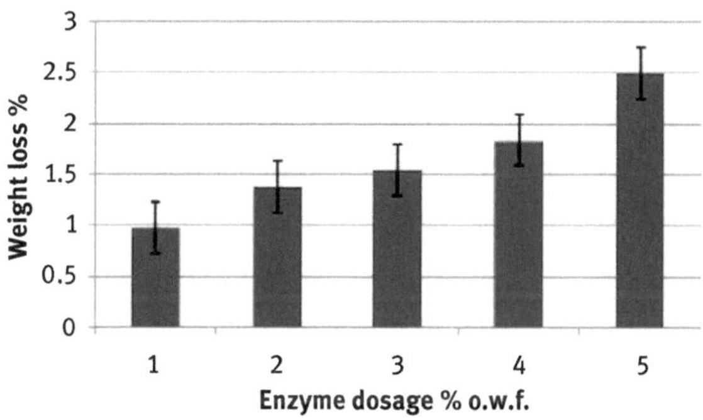

Fig. 1.14: Weight loss of the samples treated with different concentrations of cellulase (o.w.f.) after 45 min treatment at the optimum conditions for enzyme activity.

The sonochemical mechanism for deposition of NPs in general, and antibacterial NPs in particular, has been already described in this chapter. A simplified mechanism for embedding ZnO NPs onto the cellulase treated cotton fibers is illustrated in Figure 1.15.

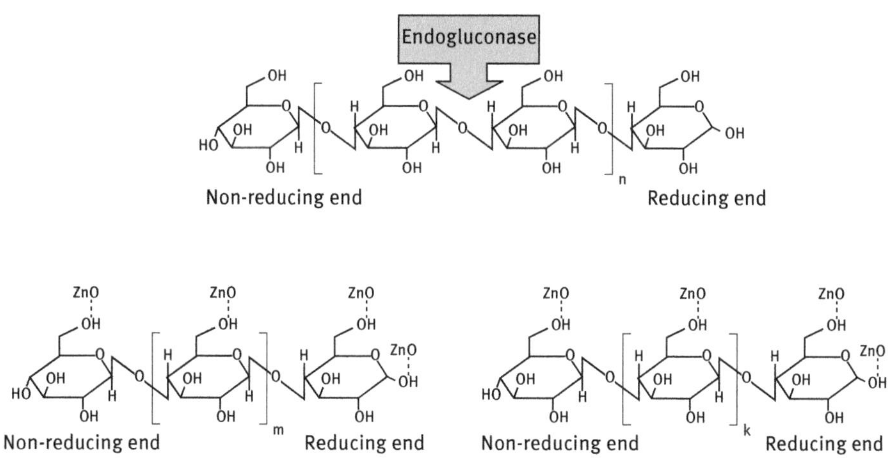

Fig. 1.15: Mechanism of interaction of ZnO NPs with cellulase pretreated cotton fibers.

The cellulase pretreatment generates numerous nucleation centers where ultra-small, uniformly distributed NPs can be formed along the fibers. The increase in free hydroxyl groups is in good agreement with the results obtained from the SEM examination, where smaller particle size was observed for the cellulase pretreated fabrics. The ZnO NPs are anchored strongly to the fabric as evidenced by the absence of NPs in the leachate from the washing of the fabrics.

The biocidal activity of ZnO-coated fabrics (treated and nontreated with enzyme) was tested against five different strains of bacteria in two modes: before and after 10 washing cycles at 92 °C. The results are presented in Figure 1.16. According to the ISO standard used for the testing, the minimum level of antibacterial activity that is deemed acceptable is 2. Good activity levels, above 2, for three out of the five bacterial species assessed were obtained for ZnO fabric nontreated with enzyme. The antibacterial activity levels against *P. aeruginosa* and *A. baumannii* that commonly display intrinsic resistance to many antibiotics and antimicrobial compounds were below 2 [75, 76].

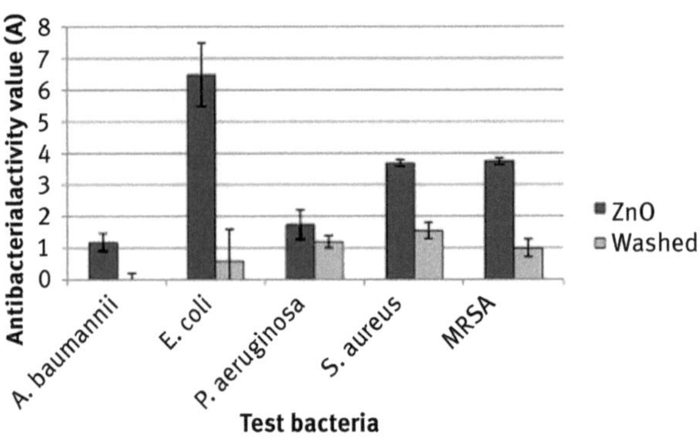

Fig. 1.16: Antibacterial activity values for ZnO coated cotton before and after 10 washing cycles at 92 °C. Antibacterial efficacy was evaluated according to the absorption method described in ISO 20743:2007. Error bars indicate standard deviation (n = 3).

In all cases, a reduction of the antibacterial activity values was observed after 10 washing cycles. This reduction was presumably due to the gradual dissolution of ZnO NPs from the fibers during repeated washing. Partial dissolution of ZnO is governed by the K_{sp} of 10^{-9}. Intact NPs were not found in the washing solution, but a small amount of Zn^{+2} was detected by ICP.

The antibacterial efficacy levels for the enzyme pretreated ZnO-coated cotton fabric against the same five species of bacteria are shown in Figure 1.17. The antibacterial activity levels before washing the fabrics were higher than 2 against all the bacterial species tested. Although the antibacterial activity again decreased after washing, the levels for 3 out of 5 bacterial species remained well above the minimum acceptable level.

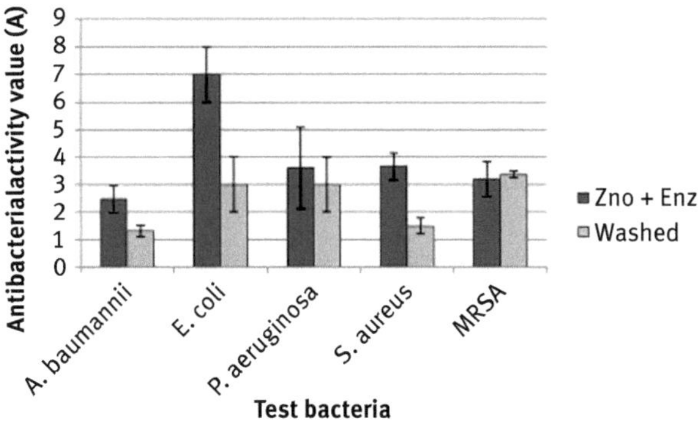

Fig. 1.17: Antibacterial activity values for enzyme treated ZnO-coated cotton before and after 10 wash cycles at 92 °C. Antibacterial efficacy was evaluated according to the absorption method described in ISO 20743:2007. Error bars indicate standard deviation (n = 3).

The amount of ZnO was very similar on the untreated and enzyme pretreated fabrics (0.75 and 0.7 % wt respectively); and for *S. aureus*, *E. coli*, and *MRSA* there was no significant difference in the antibacterial activity between the fabrics. In each of these cases the levels of activity observed were quite high, indicating that these strains were susceptible to the killing effects of the ZnO NPs both on the pretreated and untreated cotton. For *P. aeruginosa* and *A. baunmannii* a better level of activity was observed with the cellulase pretreated fabric. There was some small degree of antibacterial activity associated with the nonenzyme treated ZnO-coated cotton, but the smaller and more evenly distributed ZnO NPs on the enzyme pre-treated cotton appeared to work more effectively.

The main observable morphological change of the coating associated with the enzymatic pretreatment was size reduction of the NPs (Fig. 1.12). An improvement in the antibacterial activity resulting from a decrease in NP size has been reported previously [62].

Summarizing the results on the enzymatic method for treatment of cotton it can be concluded that biopolishing reduced the surface hairiness and improved the surface evenness of the cotton fibers. An increased number of reducing ends in the cellulose structure as a consequence of the cellulase treatment improved the ZnO NPs' adhesion due to increasing hydrogen bonding with the fabric. The smaller size of the NPs translated into an improved level of antibacterial activity against some species of bacteria. The positive effect of the pretreatment on stability of the ZnO coatings and their antibacterial activity after repeated washing at 92 °C was demonstrated.

1.3.5 Size dependence of the antibacterial activity of ZnO NPs

The effect of particle size on the antibacterial properties of metal oxide NPs is a subject of interest for many researchers. Starting from the works of Sawai [60, 61] and Yamamoto [55] it was undoubtedly confirmed that a decrease in particle size enhances the biocidal activity of metal nano-oxides.

The size of nanoparticles may have a greater impact on their activity, probably because of a greater accumulation of the nanoparticles inside the cell membrane and cytoplasm [77]. It was indicated that ZnO NPs at a concentration of between 3 and 10 mM caused 100 % inhibition of bacterial growth as a result of the intracellular accumulation of nanoparticles [78]. Stoimenov *et al.* suggested that electrostatic interactions between the bacteria surface and nanoparticles could be involved in the mechanism [79]. A small amount of physical damage to the cell envelope might also be induced by ZnO [80].

The experimental evidence from the literature demonstrates that the dominant mechanisms of the observed antibacterial behavior of ZnO particles is the generation of radical oxygen species (ROS) and the resulting interaction between ROS and the cell [60, 61, 81, 82]. Recent studies by Northern analyses of various reactive oxygen species (ROS) specific genes and confocal microscopy suggest that the antibacterial activity of ZnO nanoparticles might involve both the production of reactive oxygen species and the accumulation of nanoparticles in the cytoplasm or on the outer membranes [83].

This fact that antibacterial formulations of nanoparticles themselves may be effective in controlling the outbreak of new resistant strains of bacteria encourages the development of new nanoparticles-based composites that are stable, robust and durable, and that are effective in the destruction or elimination of bacteria and viruses. From this point of view the sonochemical method that gives an opportunity to produce composites of metal oxide NPs with various substrates hardly can be overestimated. In our work it was demonstrated that metal oxide NPs deposited on textiles are very effective antibacterial materials. Ecologically friendly one-stage ultrasound assisted deposition provides strong attachment of active nano-oxides to the substrate and long duration of their biocidal function. Optimization of process conditions like time, temperature and concentration of precursor will result in the creation of coatings with designed size and properties.

1.4 Conclusion

Nanoparticles with intrinsic antibacterial properties (ZnO, CuO and TiO_2) could be synthesized and uniformly deposited onto the surface of different kinds of textiles by the sonochemical method. The coating was performed by a simple, efficient, one-step procedure using environmentally friendly reagents. The physical and chemical analyses demonstrated that nanocrystals are finely dispersed onto the fabric surfaces without any significant damage of the structure of yarns. The mechanism of nanoparticle formation and adhesion to the textile was discussed. It is based on the local melting of the substrate due to the high rate and temperature of nanoparticles thrown at the solid surface by sonochemical microjets. The strong adhesion of metal nano-oxides to the substrate was demonstrated in terms of absence of leaching of the NPs into the washing solution. The performance of fabrics coated with low content of active nanomaterial ($< 1\,wt\%$) as an antibacterial agent was investigated and their excellent bactericidal effect was demonstrated.

Coated fabrics have potential applications in wound dressings, bed linings and as bandages. They can also be recommended for the purification of medical and food equipment, domestic cleaning, *etc.*

Bibliography

[1] Petrulyte S. Advanced textile materials and biopolymers in wound management. Dan Med Bull 2008; 55: 72–7.
[2] Gao Y, Cranston R. Recent advances in antimicrobial treatments of textiles. Textile Res J 2008; 78: 60–72.
[3] Craighead H, Leong K. Nanotechnology research direction, in: Biotechnology, Medicine, and Healthcare. The Netherlands: Kluwer Academic Publishers, 2000: 153–6.
[4] Qian L, Hinestroza JP. Application of nanotechnology for high performance textile. J Textile Apparel Technol Management 2004; 4: 1–7.
[5] Wong YWH, Yuen CWM, Leung MYS, Ku SKA, Lam HLI. Selected applications of nanotechnology in textiles. AUTEX Research J 2006; 6: 1–8.
[6] Coyle S, Wu Y, Lau KT, De Rossi D, Wallace G, Diamond D. Smart nanotextiles. MRS Bulletin 2007; 32: 434–7.
[7] Klueh U, Wagner V, Kelly S, Johnson A, Bryers JD. Efficacy of silver-coated fabric to prevent bacterial colonization and subsequent device-based biofilm formation. J Biomed Mater Res 2000; 53: 621–31.
[8] Lee HJ, Yeo SY, Jeong SH. Antibacterial effect of nanosized silver colloidal solution on textile fabrics. J Mat Sci 2003; 38: 2199–204.
[9] Gorenšek M, Recelj P. Nanosilver functionalized cotton fabric. Textile Res J 2007; 77: 138–41.
[10] Searle A. The Use of Metal Colloids in Health and Disease. Sutton: N.Y., 1919.
[11] Birringer R. Nanocrystalline materials. Mater Sci Eng A 1989; 117: 33–43.
[12] Shukla S, Seal S, Schwartz S, Zhou D. Synthesis and characterization of nanocrystalline silver coating of fly ash cenosphere particles by electroless process. J Nanosci Nanotechnol 2001; 417–24.
[13] Richard P, LeFloch R, Chamoux C, Pannie M, Espaze E, Richet H. Pseudomonas aeruginosa outbreak in a burn unit: role of antimicrobials in the emergence of multiply resistant strains. J Inf Dis 1994; 170: 377–83.
[14] Wendler F, Meister F, Montigny R, Wagener M. A new antimicrobial ALCERU fiber with silver nanoparticles. Fibers and Tex in East Europe 2007; 15: 64–5.
[15] Duran N, Marcato PD, De Souza GIH, Alves OL, Esposito E. Antibacterial effect of silver nanoparticles produced by fungal process on textile fabrics and their effluent treatment. J Biomed Nanotechnol 2007; 13: 203–8.

[16] Yuranova T, Rincon AG, Pulgarin C et al. Performance and characterization of Ag-cotton and Ag/TiO_2 loaded textiles during the abatement of E. coli. J Photochem Photobiol A 2006; 181: 363–9.
[17] Scholz J, Nocke G, Hollstein F, Weissbach A. Investigations on fabrics coated with precious metals using the magnetron sputter technique with regard to their anti-microbial properties. Surf Coat Technol 2005; 192: 252–6.
[18] Daoud WA, Xin JH. Low temperature sol-gel processed photocatalytic titania coating. J Sol-Gel Sci Tech 2004; 29: 25–9.
[19] Uddin F. Some aspects in textile nanofinishing. 86thTextile-Institute World Conf., Conf. Proceed. 2: U291–307. Hong Kong: Polytechnic Univ, Inst Textiles and Clothing Press, 2008.
[20] Sojka-Ledakowicz J, Lewartowska J, Kudzin M, Jesionowski T, Siwińska-Stefańska K, Krysztafkiewicz A. Modification of textile materials with micro-and nano-structural metal oxides. Fibers and Textiles in Eastern Europe A 2008; 16: 112–6.
[21] Shi L, Zhao Y, Zhang X, Su H, Tan H. Antibacterial and anti-mildew behavior of chitosan/nano-TiO_2 composite emulsion. Korean J Chem Eng 2008; 25: 1434–8.
[22] El-Naggar AM, Zohdy MH, Hassan MS, Khalil EM. Antimicrobial protection of cotton and cotton/polyester fabrics by radiation and thermal treatments. I. Effect of ZnO formulation on the mechanical and dyeing properties. J Appl Polym Sci 2003; 88: 1129–37.
[23] Zohdy MH, Kareem HA, El-Naggar AM, Hassan MS. Microbial detection, surface morphology, and thermal stability of cotton and cotton/polyester fabrics treated with antimicrobial formulations by a radiation method J Appl Polym Sci 2003; 89: 2604–10.
[24] Schollmeyer, E. Nanotechnology for functionalization of textile materials. 3rd International Textile Clothing and Design Conference, 2007, Dubrovnik,Croatia, Tekstil 2007; 56: 75–8.
[25] Xue CH, Jia ST, Chen HZ, Wang M. Preparation of superhydrophobic surfaces on cotton textiles. Sci Technol Adv Mater 2008; 9: 035001-7.
[26] Wang R, Xin JH, Tao XM, Daoud WA. ZnO nanorods grown on cotton fabrics at low temperature. Chem Phys Lett 2004; 398: 250–5.
[27] Daoud WA, Xin JH, Zhang YH. Surface functionalization of cellulose fibers with titanium dioxide nanoparticles and their combined bactericidal activities. Surf Sci 2005; 599: 69–75.
[28] Qi, KH, Chen XQ, Liu YY. Facile preparation of anatase/SiO_2 spherical nanocomposites and their application in self-cleaning textiles. J Mater Chem 2007; 17: 3504–8.
[29] Haufe H, Muschter K, Siegert J, Bottcher H. Bioactive textiles by sol-gel immobilised natural active agents J Sol-Gel Sci Technol 2008; 45: 97–101.
[30] Vigneshwaran N, Kumar S, Kathe AA, Varadarajan PV, Prasad V. Functional finishing of cotton fabrics using zinc oxide-soluble starch nanocomposites. Nanotechnology 2006; 17: 5087–95.
[31] Xue CH, Jia ST, Zhang J, Ma JZ. Large-area fabrication of superhydrophobic surfaces for practical applications: an overview. Sci Technol Adv Mater 201; 11: 033002 (15pp).
[32] Mao Z, Shi Q, Zhang L, Cao H. The formation and UV-blocking property of needle-shaped ZnO nanorod on cotton fabric. Thin Solid Films 2006; 517: 2681–6.
[33] Gabbay J, Borkow G, Mishal J, Magen E, Zatcoff R, Shemer-Avn, Y. Copper oxide impregnated textiles with potent biocidal activities. J Ind Textiles 2006; 35: 323–35.
[34] Borkow GM, Gabbay J. Copper as a biocidal tool. Current Med Chem 2005; 12: 2163–75.
[35] Suslick KS, Price GJ. Application of ultrasound to materials chemistry. J Annu Rev Mater 1999; 29: 295–326.
[36] Gedanken A. Using sonochemistry for the fabrication of nanomaterials. Ultrason Sonochem 2004; 11: 47–55.
[37] Suslick KS, Choe SB, Ciclowlas AA, Grinstaff MW. Sonochemical synthesis of amorphous iron. Nature 1991; 353: 414–6.

[38] Mason TJ. Sonochemistry. London: Royal Society of Chemistry Press, 1990.
[39] Doktycz SJ, Suslick KS. Interparticles collisions driving by ultrasound. Science 1990; 247: 1067–9.
[40] Suslick KS, Hammerton DA, Cline RE. Sonochemical hot spot. J Am Chem Soc 1986; 108: 5641–2.
[41] Gedanken A. Doping nanoparticles into polymers and ceramics using ultrasound radiation. Ultrason Sonochem 2007; 14: 418–30.
[42] Mason TJ. Developments in ultrasound – non-medical. Prog Biophys Molec Biol 2007; 93: 166–75.
[43] Vajnhandl S, Le Marechal AM. Ultrasound in textile dyeing and the decolouration/mineralization of textile dyes. Dyes and Pigm 2005; 65: 89–101.
[44] Perelshtein I, Applerot G, Perkas N et al. CuO – cotton nanocomposite: Formation, morphology, and antibacterial activity. Surf Coat Technol 2009; 204: 54–7.
[45] Perelshtein I, Applerot G, Perkas N, Grinblat J, Gedanken A. A one-step process for the antimicrobial finishing of textiles with crystalline TiO_2 nanoparticles. Chemistry – A European Journal 2012; 18: 4575–80.
[46] Kim SJ, Lee EG, Park SD et al. Photocatalytic effects of rutile phase TiO_2 ultrafine powder with high specific surface area obtained by a homogeneous precipitation process at low temperatures. J Sol-Gel Sci Technol 2001; 22: 63–74.
[47] Tokunaga Y, Uchiyama H, Oaki Y, Imai H. Specific photocatalytic performance of nanostructured rutile-type TiO_2: selective oxidation of thiazine dye with a bundled architecture consisting of oriented nanoneedles. Sci Adv Mater 2010; 2: 69–73.
[48] Caballero L, Whitehead KA, Allen NS, Verran J. Inactivation of Escherichia coli on immobilized TiO_2 using fluorescent light. J Photochem Photobiol A: Chem. 2009; 202: 92–8.
[49] Kim b, Kim D, Cho D, Cho S. Bactericidal effect of TiO_2 photocatalyst on selected food-borne pathogenic bacteria. Chemosphere 2003; 52: 277–81.
[50] Madrid PA, Moorillon GVN, Borunda EO, Yoshida MM. Photoinduced bactericidal activity against Pseudomonas aeruginosa by TiO_2 based thin films. FEMS Microbiol Lett 2002; 211: 183–8.
[51] Kangwansupamonkon W, Lauruengtana V, Surassmo S, Ruktanonchai, U. Antibacterial effect of apatite-coated titanium dioxide for textiles applications Nanomedicine: Nanotechnology 2009; 5: 240–9.
[52] Wu D, Long M, Zhou J et al. Synthesis and characterization of self-cleaning cotton fabrics modified by TiO_2 through a facile approach. Surf Coat Technol 2009; 203: 3728–33.
[53] Konovalova TA, Lawrence J, Kispert LD. Generation of superoxide anion and most likely singlet oxygen in irradiated TiO_2 nanoparticles modified by carotenoids. J Photochem Photobiol A: Chem 2004; 162: 1–8.
[54] Liu GY, Essex A, Buchanan JT et al. Staphylococcus aureus golden pigment impairs neutrophil killing and promotes virulence through its antioxidant activity. J Exp Med 2005; 202: 209–15.
[55] Yamamoto O. Influence of particle size on the antibacterial activity of zinc oxide. Int Inorg Mater 2001; 3: 643–6.
[56] Perelshtein I, Applerot G, Perkas N et al. Antibacterial properties of an in situ generated and simultaneously deposited nanocrystalline ZnO on fabrics. ACS Appl Mater Interfaces 2009; 1: 361–6.
[57] Li N, Xia T, Nel ME. The role of oxidative stress in ambient particulate matter-induced lung diseases and its implications in the toxicity of engineered nanoparticles. Free Radic Biol Med 2008; 44: 1689–99.

[58] Neal AL. What can be inferred from bacterium – nanoparticle interactions about the potential consequences of environmental exposure to nanoparticles? Ecotoxicology 2008; 17: 362–71.
[59] Hu X, Cook S, Wang P, Hwang H. In vitro evaluation of cytotoxicity of engineered metal oxide nanoparticles. Sci Total Environ 2009; 407: 3070–2.
[60] Sawai J, Igarashi H, Hashimoto A, Kokugan T, Shimizu M. Effect of particle size and heating temperature of ceramic powders on antibacterial activity of their slurries. J Chem Eng Jpn 1996; 29: 251–6.
[61] Sawai J, Shoji S, Igarashi H et al. Hydrogen peroxide as an antibacterial factor in zinc oxide powder slurry. J Ferment Bioeng 1998; 86: 521–2.
[62] Applerot G, Lipovsky A, Dror R et al. Enhanced antibacterial activity of nanocrystalline ZnO due to increased ROS-mediated cell injury. Adv Funct Mater 2009; 19: 842–52.
[63] Soderberg T, Agren M, Tengrup I, Hallmans G, Banck G. The effects of an occlusive zinc medicated dressing on the bacterial flora in excised wounds in the rat. Infection 1989; 17: 81–5.
[64] Lansdown ABG, Mirastschijski U, Stubbs N, Scanlon E, Agren MS. Zinc in wound healing: Theoretical, experimental, and clinical aspects. Wound Rep Reg 2007; 15: 16–22.
[65] Sengupta G, Ahluwalia HS, Banerjee S, Sen SP. Chemisorption of water-vapor on zinc oxide. J Colloids Interface Sci 1979; 69: 217–24.
[66] Kassenbeck P. Bilateral structure of cotton structure as revealed by enzymatic degradation. Textile Res J 1970; 40: 330–340.
[67] Li Y, Hardin IR. Enzymatic scouring of cotton: effects on structure and properties. Text Chem Color 1997; 29: 71–7.
[68] Cavaco-Paulo A. Mechanism of cellulase action in textile processes. Carbohydrate Polym 1998; 37: 273–7.
[69] Polaina J, MacCabe AP. Industrial Enzymes: Structure, Function and Applications. Springer, Dordrecht, 2007.
[70] Ibrahim N, Fahmy H, Hassan T, Mohamed Z. Effect of cellulase treatment on the extent of post-finishing and dyeing of cotton fabrics. J Mater Proces Technol 2005; 160: 99–106.
[71] Yang CQ, Zhou W, Lickfield GC, Parachura K. Cellulase treatment of durable press finished cotton fabric: effects on fabric strength, abrasion resistance, and handle. Textile Res J 2003; 73: 1057–62.
[72] Bodalo A, Gomez JL, Gomez E, Maximo F, Montiel MC. Kinetics of the L-aminoacylase-catalyzed resolution of N-acetyl-DL-butyrine. J Chem Technol Biotechnol 2002; 77: 552–8.
[73] Perelshtein I, Ruderman Y, Perkas N et al. Enzymatic pretreatment as a means of enhancing the antibacterial activity and stability of ZnO nanoparticles sonochemically coated on cotton fabrics. J mater Chem 2012; 22: 10736–41.
[74] Abramov OV, Gedanken A, Koltypin Y et al. Pilot scale sonochemical coating of nanoparticles onto textiles to produce biocidal fabrics. Surf Coat Tech 2009; 204: 718–22.
[75] Lambert A. Mechanisms of antibiotic resistance in Pseudomonas aeruginosa. J R Soc Med 2002; 95: 22–6.
[76] Coyne S, Courvalin P, Périchon B. Efflux-mediated antibiotic resistance in acinetobacter spp. J Antimicrob Agents Chemother: Antim Agen and Chem 2001; 55: 947–53.
[77] Jones N, Ray B, Ranjit KT, Manna AC. Antibacterial activity of ZnO nanoparticles suspensions on a broad spectrum of microorganisms. FEMS Microbiol Lett 2008; 279: 71–6.
[78] Brayner R, Ferrari-Iliou R, Brivois N et al. Toxicological impact studies based on Escherichia coli bacteria in ultrafine ZnO nanoparticles colloidal medium. Nano Lett 2006; 6: 866–70.
[79] Stoimenov PK, Klinger RL, Marchin GL, Klabunde KJ. Metal oxide nanoparticles as bactericidal agents. Langmuir 2002; 18: 6679–86.

[80] Zhang L, Jiang Y, Ding Y, Povey M, York D. Investigation into the antibacterial behaviour of suspensions of ZnO nanoparticles (ZnO nanofluids). J Nanoparticles Res 2007; 9: 479–89.
[81] Lipovsky A, Tzitrinovich Z, Friedmann H, Applerot G, Gedanken A, Lubart R. EPR study of visible light-induced ROS generation by nanoparticles of ZnO. J Phys Chem C 2009; 113: 15997–16001.
[82] Zhang L, Jiang Y, Ding Y et al. Mechanistic investigation into antibacterial behavior of suspensions of ZnO nanoparticles against E. coli. J Nanopart Res 2010; 12: 1625–36.
[83] Raghupathi KR, Koodali RT, Manna AC. Size-dependent bacterial growth. Inhibition and mechanism of antibacterial activity of zinc oxide nanoparticles. Langmuir 2011; 27: 4020–8.

Dario Brancaleoni and Federico Squassabia, with the technical support of Zinsser Analytic GmbH

2 Automated solutions for high-throughput experimentation in heterogeneous catalyst research

2.1 Introduction

The pharmaceutical industry has been using high-throughput synthesis and screening for many years and automation tools are well known. The current requirements from the petrochemical industry to produce small quantities of catalysts for screening purposes are raising new automation challenges. To address these needs for high-throughput experimentation in catalyst research, Zinsser Analytic has developed innovative automation tools.

Zinsser Analytic, a German company located in Frankfurt, has wide experience in the automation field in the following technologies in particular:

- **Automatic Liquid Handling** (pipetting and sample preparation systems since 1979),
- **CombiChem** first in automation of parallel synthesis since 1988, solid and solution phase synthesis (SOPHAS), peptide synthesis (PepSy) microwave synthesis (Navigator), synthesis preparation (REDI, Calli) synthesis work-up (Speedy, Calli)
- **Screening** HIV screening platforms since 1986, assay plate preparation and reformatting systems for HT-screening (MOSS) Cherry picking (Lissy 2002, Calli 2002), CRISSY polymorph and salt screening system, SuSy for formulation and solubility studies and screening, Formula X for formulation testing and development, SOPHAS-CAT HT Factory.

In many years of development, Zinsser Analytic has found the right solution or, better, the right tool for the automation of a specific application: liquid and powder handling, pick and place functions, mixing, heating, cooling, weighing, reactors controls and inspection functions *etc.*

Defining the right system for a specific application is a long process involving many steps and both sides learning many things during the process. Having meetings with the user that will work on the project ensures that there are no misunderstandings and helps define the best solutions. The hardware engineers and programmers can also add valuable input to the design based on their vast experience from working on many other diverse projects. Bringing everybody together for meetings increases the productivity of the project and increases everybody's understanding of the final goals. Monthly update meetings help keep the project on track.

Since 1979 automated liquid handling systems with customized software have been installed in laboratories all over the world.

Today, Zinsser specializes in developing, producing and distributing flexible automated systems for combinatorial chemistry and tools for drug discovery. For many years, Zinsser Analytic has supplied customer with individual solutions.

More than 200 tools (Fig. 2.1) and modules are available in relation to the requirements of the users' applications. In the following Figure 2.1 some of these tools are shown.

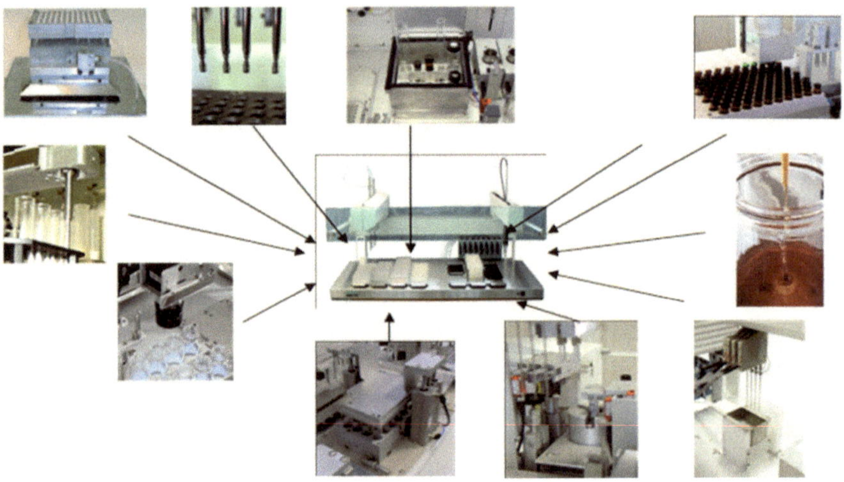

Fig. 2.1: Zinsser tools examples for different applications.

All the tools are controlled by WINLISSY SOFTWARE: it is a powerful but flexible software that permits the access of all liquid and powder handling parameters, simulation of methods, scheduling, complete audit trail, download and upload of Excel® -Files (Fig. 2.2).

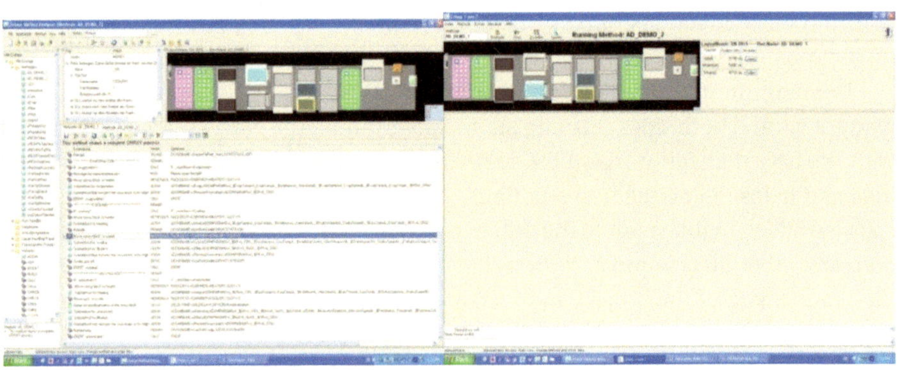

Fig. 2.2: The Winlissy software with layout and workflow.

2.2 The preparation of solid catalysts

The preparation of solid catalyst needs many steps which need specific tools. In Table 2.1 the catalyst preparation workflow is described with the right tools needed to carry out all the process in automation.

Tab. 2.1: The catalyst preparation workflow and the right tools.

Catalyst Preparation Workflow	Tools
Planning the screen	DoE-Software or Excel®
Solid dispensing	Powder Dispenser
Weight check	Balance
Distribution of solvents, reagents	Precise Liquid Handling
Mixing	Mixer
Heating / Temperature control	Heating, Cooling
pH-Monitoring and Adjustment	pH-Monitoring / Titration
Filtration	Filtration Device

The operations needed for the preparation of heterogeneous catalysts are described in Figure 2.3.

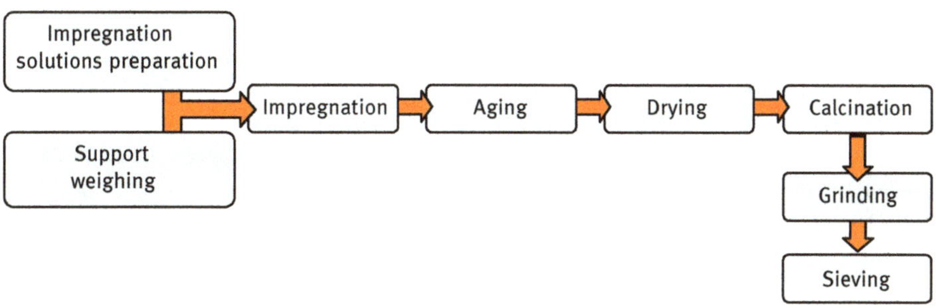

Fig. 2.3: Typical heterogeneous catalyst synthesis workflow.

2.3 Automation challenges examples

Achieving homogeneous automated impregnation: Manual processes typically deal with amounts in the range of 100 g of catalyst where screening only requires a few grams for activity testing. The manual practice to pour the impregnation solution onto the support in a 500 mL bottle and perform the impregnation by tilting does not work when only 2–3 g of support is prepared (the support bed weight is not sufficient to fall).

A common approach for small-scale impregnation is to spray the impregnation solution on the support while vortexing. Even though this technique provides good

Fig. 2.4: Homogeneous impregnation.

results in some cases, it appears not to be efficient enough to reach homogeneous impregnation (Fig 2.4) in many others. In essence, spraying causes solvent evaporation and thus inaccurate impregnation volume which is a key factor for successful impregnation. Step dispensing injection with a standard probe avoids this problem. Using only a vortexer to get the solution to penetrate into the support pores requires long vortexing times that favor attrition. In addition, the vortexing effect is to move the entire support bed where the goal is to get the solution to migrate into the pores. This is efficiently achieved with a high-frequency vibrating device that provides energy, comparable to micro waves, to the solution.

2.3.1 Integration of commercially available devices:

Some of the devices commonly used for catalyst synthesis are not designed for automation. This is for instance the case with the dessicator or muffle furnace. Integration of these devices in an automated workstation requires to automate the doors' opening and to use a 6-axis robotic arm to mimic the technician operations (Fig. 2.5)

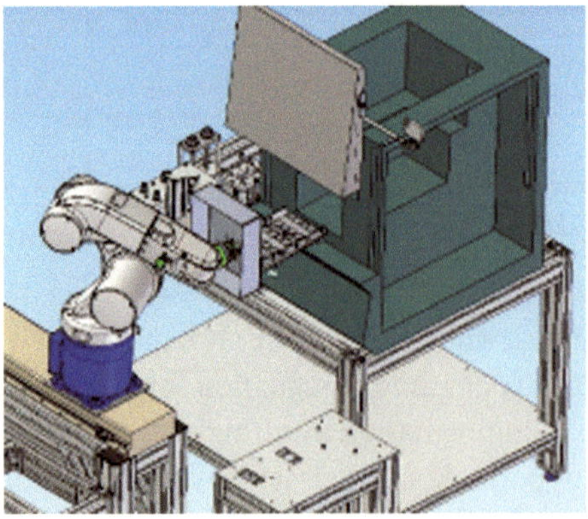

Fig. 2.5: Automated workstation.

2.4 A fully-automated solution

These automation solutions have been integrated to provide a complete workstation that operates unattended to synthesize catalysts [1, 2] ready for characterization and activity testing. The configuration is modular to suit the needs of research groups willing to increase their productivity. In particular, the system has been created with the following hardware modules: loading and unloading shelf, workbench system for synthesis, workbench system for impregnation, weighing stations for extrudates, calcination system, scraping station, pelletizer, grinding station, sieving system, 6-axis robot on 2 m track, walk-in enclosure, battery backup (UPS), safety system (Fig. 2.6).

2.4.1 SOPHAS-CAT HT

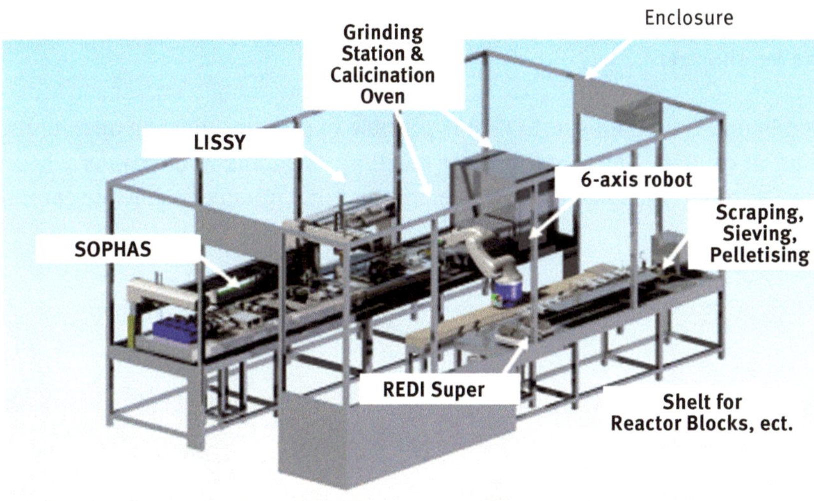

Fig. 2.6: SOPHAS-CAT HT general configuration.

2.4.2 The loading

It has created access for the user from outside, for the robot from inside. There are defined positions for reactors, sealing covers, grinding mills, sieves *etc.* and the presence of a barcode reader allows the identification of all the reagents/products on the platform. The software controls loading and positioning (Fig. 2.7).

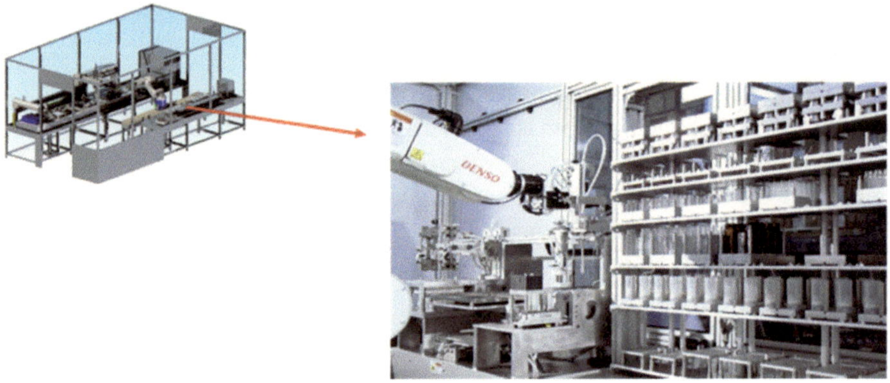

Fig. 2.7: Loading and unloading reactor blocks/plates.

2.4.3 The synthesizer

In the workbench system for synthesis it is possible to run the following operations: preparation of catalyst, liquid and powder handling, weighing in of powders, heating, mixing, stirring, reaction temperature monitoring and pH-adjustment and control (Fig. 2.8).

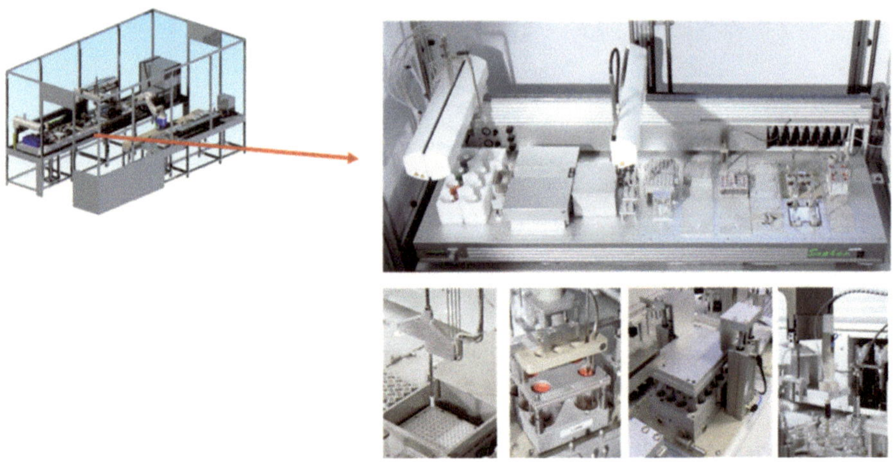

Fig. 2.8: Synthesis workbench with reactor blocks.

2.4.4 Extrudate preparation

In this platform area there is a precise dosing of extrudates: it can, but does not have to be a parallel process during the synthesis (Fig. 2.9).

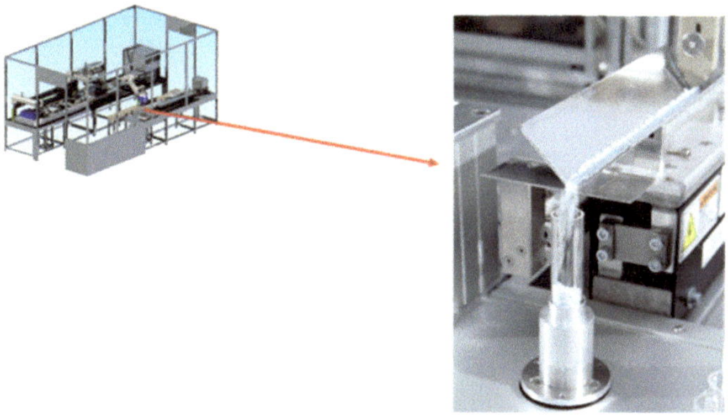

Fig. 2.9: Tool for extrudation.

2.4.5 Impregnation and drying

In this area we find the spraying tool (Zinsser spraying technology) that can manage two blocks at a time. The temperature is controlled in the parking positions and samples can be dried under nitrogen with vortexing and vacuum (Fig. 2.10).

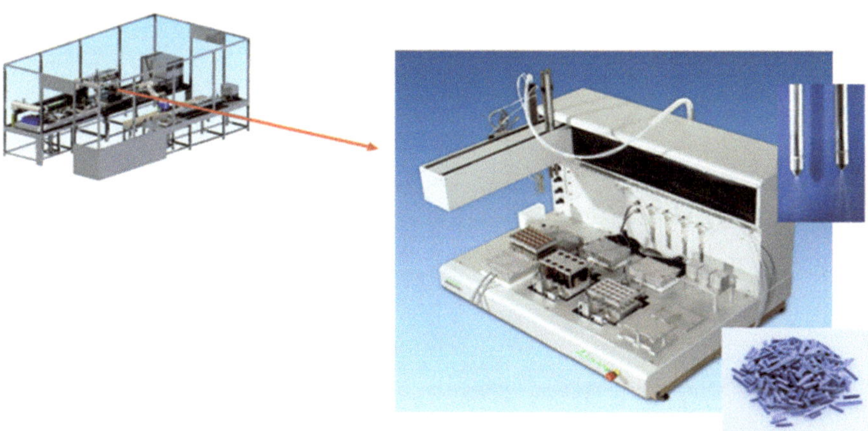

Fig. 2.10: Probes for impregnation.

2.4.6 Calcination

In this part of the system there is an integrated calcination oven with automated safety door, the loading operation is controlled by the robot. The calcination process can be carried out under nitrogen (Fig. 2.11).

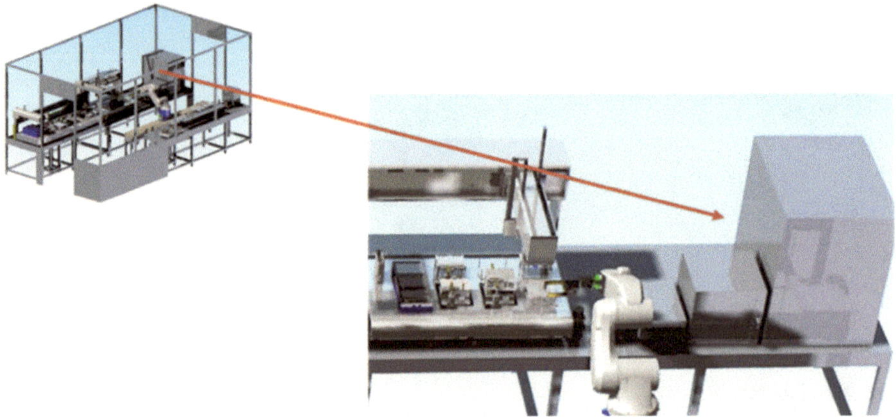

Fig. 2.11: Oven for powder calcinations.

2.4.7 Scraping and pelletizing

In this system area we find a new technology for scraping and an innovative pelletizer. The dosage of catalyst into the pelletizer is controlled strictly and there is complete automation of all the transfer operations into the grinding vessel (Fig. 2.12).

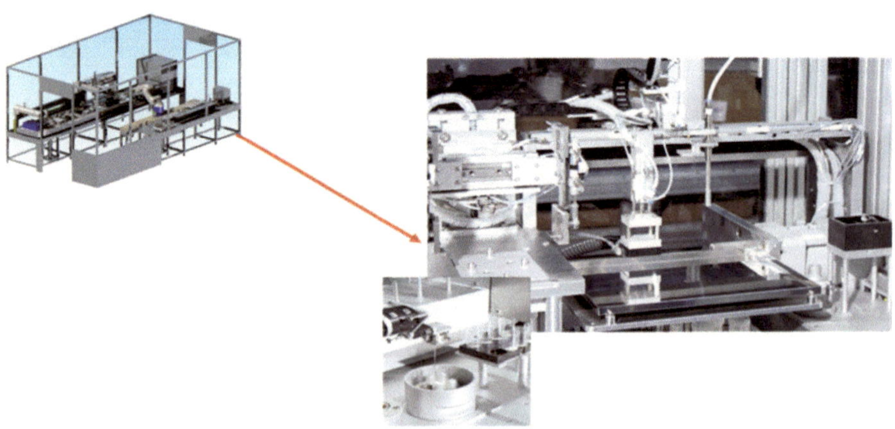

Fig. 2.12: Tools for scraping and pelletizing.

2.4.8 Grinding

The grinding station (semi-micro ball mills) can manage 4 catalysts in parallel (Fig. 2.13). There are disposable grinding vessels and balls and the operational time is very short (2–5 min).

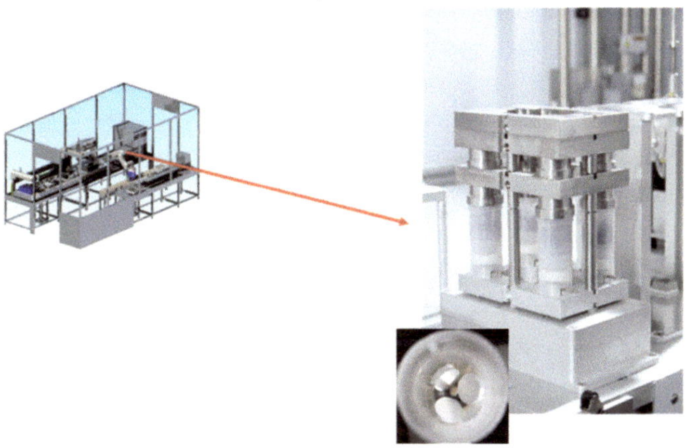

Fig. 2.13: Grinding station with 4 positions.

2.4.9 Sieving

A direct transfer from mills into the sieving station (customized mesh sizes) that can process 4 catalysts in parallel in an efficient and fast way (2–5 min). It is possible to obtain up to 3 particle sizes per catalyst and run (Fig. 2.14).

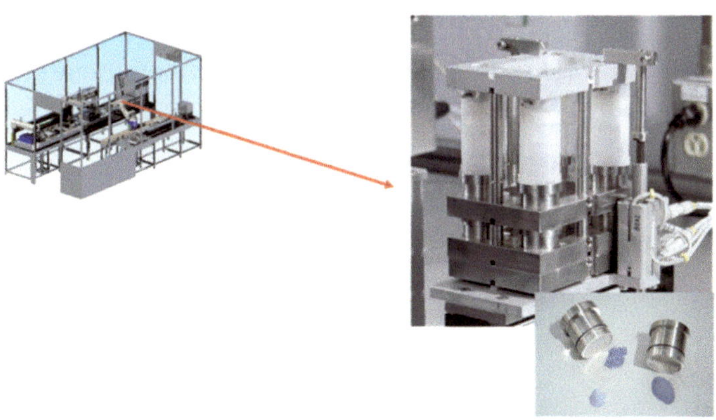

Fig. 2.14: Sieving station.

After the sieving process the catalysts are ready to be analyzed and tested; the robot transports final product to the loading bay. Software acquires data for each step and transfers them to LIMS.

2.5 Conclusion

The SOPHAS-CAT system is an automatic system for synthesis and sample preparation that comprises liquid handling, heating, cooling and mixing functions with many optional tools to aid your synthesis such as powder handling, weighing, microwave assisted synthesis, sonication and pH measurement and adjustment.

The SOPHAS-CAT (Fig. 2.15) configuration described is able to produce 200 or more sieved catalysts every 24 hours in total automation.

Fig. 2.15: SOPHAS-CAT

SOPHAS-CAT HT enables the user to run high-throughput experiments with its key features:

Easy to use: The users prepare their experiments in an Excel file and the software then tells the users which reactor blocks and reagents are required and shows them where they should be placed.

Flexible: Different experiments can be run on the same platform, programmed and optimized by the users themselves.

Modular: Should the requirements change in the future, the systems can be upgraded.

Standard labware: For the reactions, standard laboratory glass vials and tubes are used.

Customized: Smaller, less expensive systems are available as dedicated workstations for a portion of the workflow.

High-throughput experimentation in research means the possibility to manage a large number of samples with very small volumes. For this reason miniaturization helps automation reducing the cost (only milligrams) and the labor time.

Zinsser Analytic has developed in more than 30 years of experience the technologies to carry out high-throughput experimentation in automation.

Bibliography

[1] Pereira S, Clerc F, Farrusseng D, van der Waala J, Maschmeyera T, Mirodatos C. Effect of the genetic algorithm parameters on the optimisation of heterogeneous catalysts. QSAR Comb. Sci. 2005; 24.

[2] Tibiletti D, Bart de Graaf E A, Pheng Teh S, Rothenberg G, Farrusseng D, Mirodatos C. Selective CO oxidation in the presence of hydrogen: fast parallel screening and mechanistic studies on ceria-based catalysts. Journal of Catalysis 2004; 225: 489–97.

Cosimino Malitesta and Eleonora Margapoti

3 Insights from XPS on nanosized inorganic materials

3.1 Introduction

X-ray photoelectron spectroscopy (XPS) is the most used surface technique in chemical characterization of solid materials due to its speciation capability and the often easy interpretation of relevant spectra by the oxidation state concept. For this characteristic, the technique is also referred to as electron spectroscopy for chemical analysis (ESCA) coined by Kai Siegbahn, who was awarded a Nobel Prize (1981) for his development of it. Another valuable feature is represented by the low-induced damage. XPS has been used since its inception in characterization of nanosized materials (*e.g.*, catalysts) [1, 2].

However, caution should be used in the interpretation of XP spectra originated from nanosized objects, considering peculiar aspects occurring in those systems that have been recognized for a long time. In fact, size, shape of nanoparticles and the structure of their ensembles can influence both absolute and relative peak intensities and peak energies. As far as binding energy shifts are concerned, several sources have been identified and likely apply to varying degrees in different circumstances. They include changes in the particle energy levels due to size, lattice strain, charge shifting to a substrate interaction, and final state relaxation effects. Also signal background can be modified.

Even so, information on topics like contamination, particle coatings and oxidation, particle size, particle location, surface acidity, and electrical properties of particles can be obtained [3].

Excellent books are available on the fundamentals of XPS and are suggested to inexpert readers before approaching this chapter.

In the chapter peculiar effects, as well as open problems, to be considered in the nanodomain have been enlightened through significant examples of the application of XPS to nanosized inorganic materials.

3.2 XPS in the nanodomain

The simplest application of XPS to nanosized materials consists in establishing their elemental composition. In a parallel TEM and XPS study devoted to preparation of core-shell nanoparticles it was possible to reveal core consisted of $LuPO_4$ (Fig. 3.1).

In fact, XP spectra of nanoparticles showed Lu and P signals which were not detected in the control (*i.e.*, void apoferritin nanoparticles).

At a more sophisticated level of application, XPS can give information on chemical speciation and/or structures at nanometer level, because:

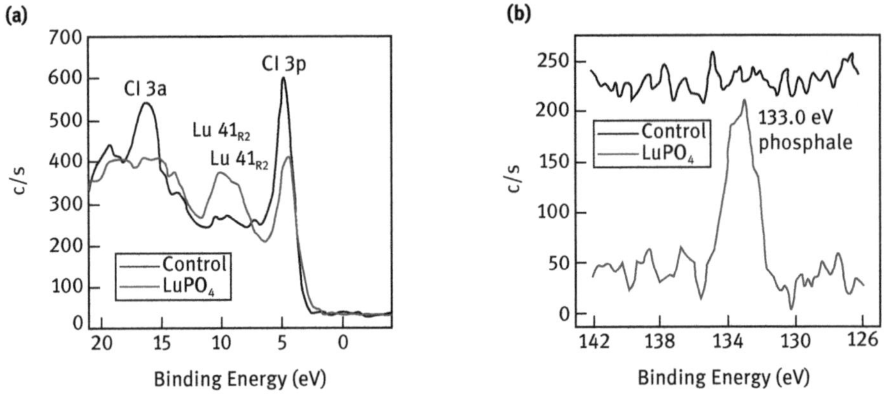

Fig. 3.1: XP spectra in (a) Lu4f and (b) P2p regions for core-shell nanoparticles with apoferritin shells and $LuPO_4$ (blue line) apoferritin nanoparticles (black line). Reproduced from [3].

- kinetic energies of photoelectrons depend on the energy states of emitting atoms and ions left behind, which are peculiar in the nanometer domain, so that information on chemical states in the sample can be drawn from energy parameters as binding energies and Auger parameters;
- photoelectrons contributing to XP peaks originate in deep layers of solid matter due to the penetrating power of employed X-rays. However, electron scattering limits information depth of XPS in the order of ten nanometers. In addition, it depends on the kinetic energy of photoelectrons and the take-off angle. As part of the emitted photoelectrons suffers from scattering, peak intensities and background arising from scattered photoelectrons contain information on sample structure at nanometer level.

In a first example of nanostructure investigation by XPS, the average size d of Cu clusters on Dow Cyclotene 3022 (BCB) was estimated by the intensity ratio of two XPS peaks of the same Cu element:

$$R = \frac{I_1^0 \left[1 - \exp\left(-\frac{d}{\lambda_1}\right)\right]}{I_2^0 \left[1 - \exp\left(-\frac{d}{\lambda_2}\right)\right]} \tag{3.1}$$

where indices 1 and 2 refer to $Cu2p_{3/2}$ and Cu 3d, respectively, of each XPS signal, λ is the inelastic mean free path of kinetic energy of the relevant signal.

The difference in probed depths of the two peaks, having different kinetic energies, allows to calculate cluster size. As a confirmation of the model a good correlation between estimated cluster size and the amount of deposited Cu could be observed (Fig. 3.2).

Another interesting feature is the possibility to have information on the shape of the nanoparticles in core-shell systems. In fact, different ratios of intensities of sig-

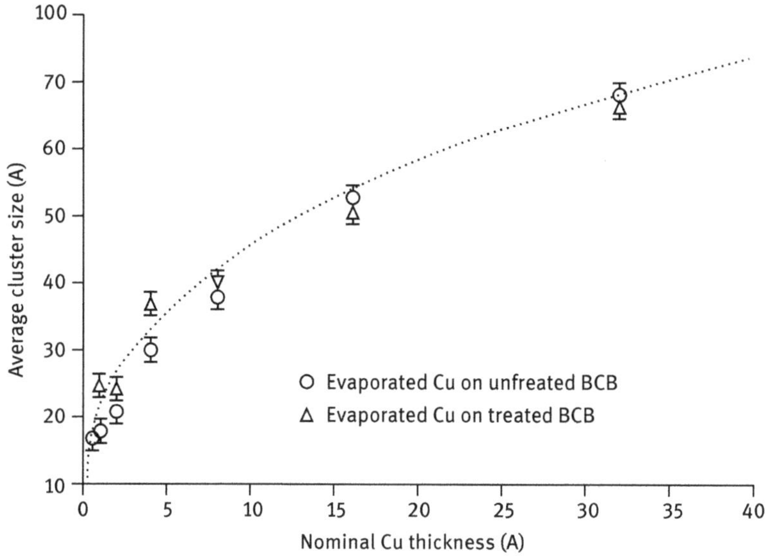

Fig. 3.2: The average size of Cu clusters as a function of the nominal thickness of evaporated Cu on BCB. Reproduced from [4].

nals from coating to substrate can be obtained for differently shaped nanoparticles (Fig. 3.3).

Even coating thickness can be measured by XPS through the possibility to perform chemical speciation. In the case of oxide coating on elemental samples, a pilot study [5] performed under the auspices of the Consultative Committee for Amount of Substance (CCQM), a body of the International Committee for Weights and Measures (CIPM), demonstrated XPS correctly applied compared well with other techniques for thickness measurements; at same time, it showed a zero offset (Tab. 3.1) which is of great significance when thicknesses in the order of nanometers occur.

Tab. 3.1: Offset values with the standard deviations for different methods applied in thickness measurements. Adapted from [5].

Method	Offset (nm)
XPS	-0.013 ± 0.110
Neutron reflectometry	0.185 ± 0.050
NRA	0.480 ± 0.122
MEIS	0.483 ± 0.108
GIXRR	0.551 ± 0.004
RBS, EDS	0.568 ± 0.263
TEM	0.804 ± 0.361
Ellipsometry	1.016 ± 0.174

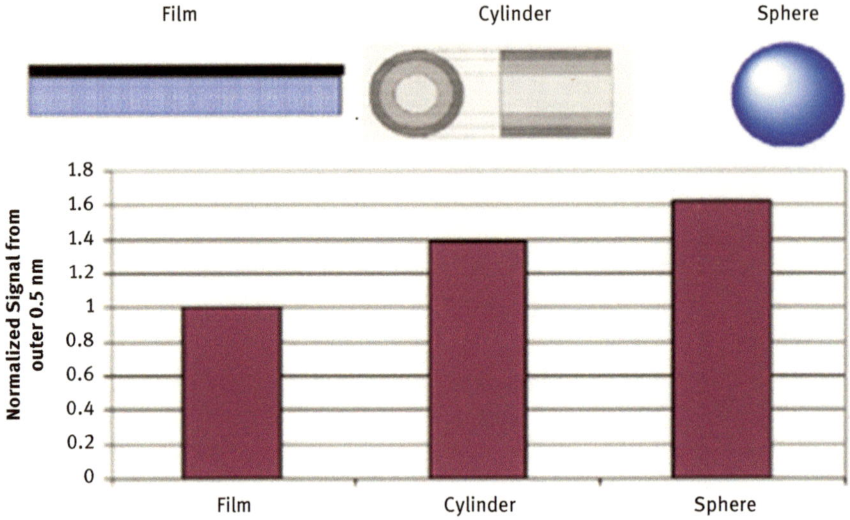

Fig. 3.3: Diagram showing the relative intensities of surface and substrate signals for a coating on a thin film (at normal emission, $\theta = 0$), a cylindrical particle, and a spherical particle. The nominal coating thickness is 0.5 nm and the calculations were performed using an inelastic mean free path of ≈ 1.65 nm. Reproduced from [3].

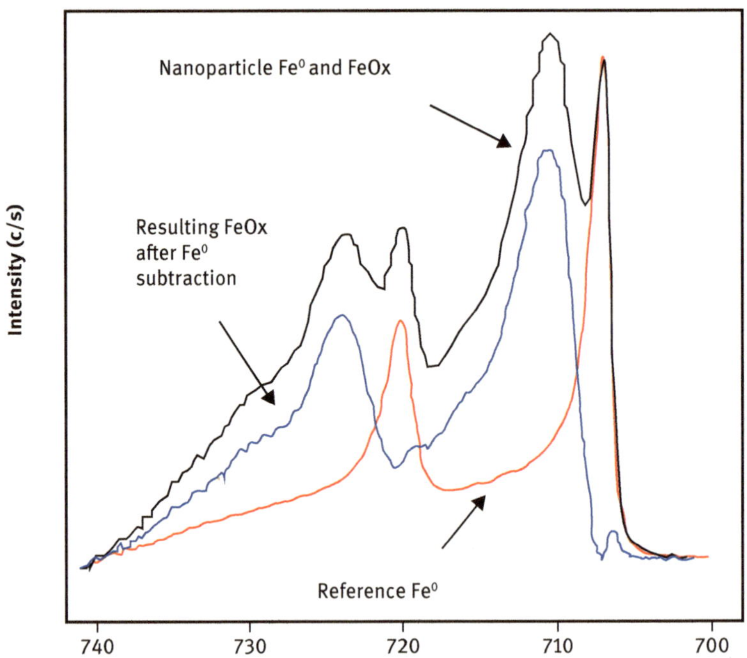

Fig. 3.4: Fe 2p photoelectron peak for an Fe nanoparticle separated into Fe metal and Fe oxide peaks. Reproduced from [3].

Applying equation (1) from [5]:

$$d = L \cos\theta \ln(1 + R_{\text{expt}}/R_0) \tag{3.2}$$

where d is the oxide thickness, L is the attenuation length of the substrate and oxide photoelectrons in the oxide, θ is the angle between electron collection direction and surface normal, R_{expt} is the ratio of the measured intensities of the photoelectrons from the oxide and the elemental states from the sample, and R_0 is the ratio of these intensities from bulk materials, it was possible to establish a 1.3 nm thickness for oxide-shell thickness in iron metal-core nanoparticles [5] (Fig. 3.4).

This estimate is not very far from the 2–3 nm value obtained by TEM/XRD, but with a possible large offset. Even the background contains information on the structure of the sample at nanometer level as it originates mainly in scattering events involving emitted photoelectrons and atoms of the samples, so being influenced by their spatial order. To extract information it is necessary to finely model the background considering the underlying physical phenomena. This was done by Tougaard, who implemented the model in the QUASES software package to be applied to XP spectra analysis. In an application, an Au 4f spectrum recorded on gold nanoclusters grown on polystyrene was analyzed (Fig. 3.5).

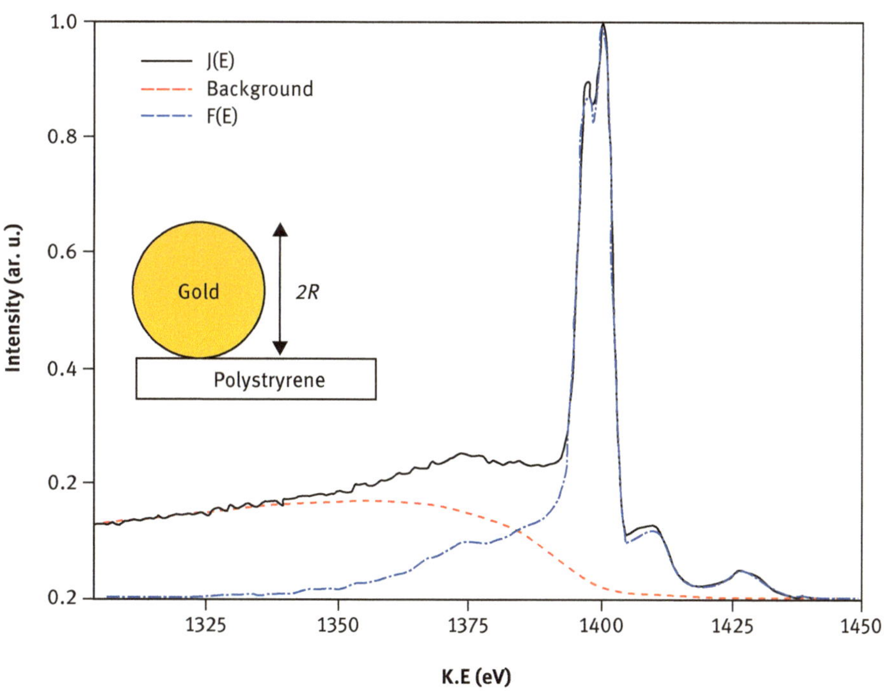

Fig. 3.5: Au 4f spectrum analyzed by QUASES package. Reproduced from [6].

By comparison with a reference spectrum, a cluster size of about 80 Å could be determined. Several authors have reported shifts of XPS signals in passing from bulk to nanomaterials. The detailed understanding of the mechanisms causing the shift is subject to ongoing research. Several effects could be considered:

- initial state effects: change in energy levels due to size (surface), lattice strain, substrate interaction, *etc.*
- final state effects: relaxation, *etc.*

Their relative importance must be evaluated in each different material, and it can influence the value and the sign of the shift. In an interesting investigation on Ni clusters on rutile TiO_2 surfaces, an energy of $Ni2p_{3/2}$ and Ni Auger peaks higher at smaller cluster size was observed in comparison to bulk material (Fig. 3.6).

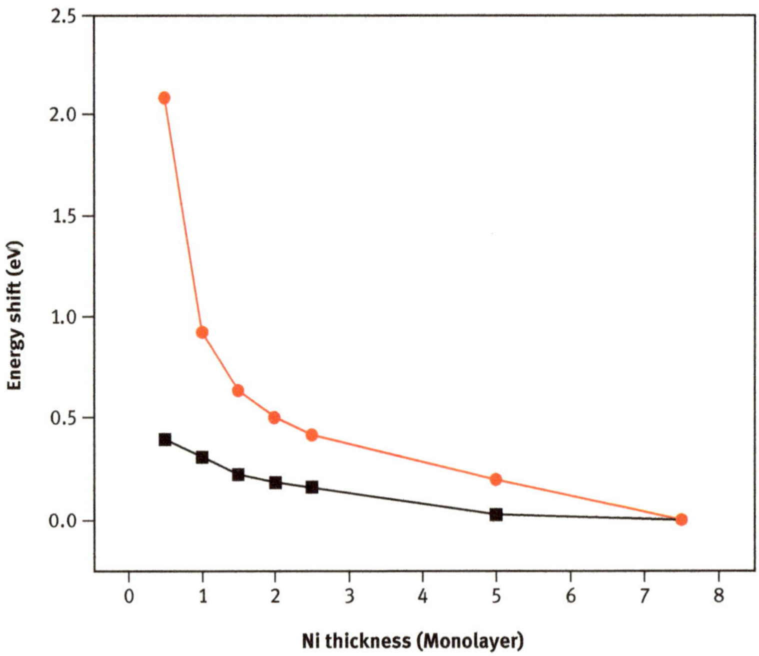

Fig. 3.6: The BE shifts of Ni 2p3/2 core-level (solid squares) and $L_3M_4,M_{4,5}$ Auger transitions (solid circles) with respect to those of bulk Ni as a function of Ni coverage on rutile TiO_2 surfaces. Reproduced from [7].

Approximating $Ni2p_{3/2}$ binding energy (BE) shift by [8]:

$$\Delta BE = -\Delta\epsilon - \Delta R$$

where ϵ is the orbital energy and R the relaxation energy, separation of initial- and final-state effects can be obtained (Fig. 3.7).

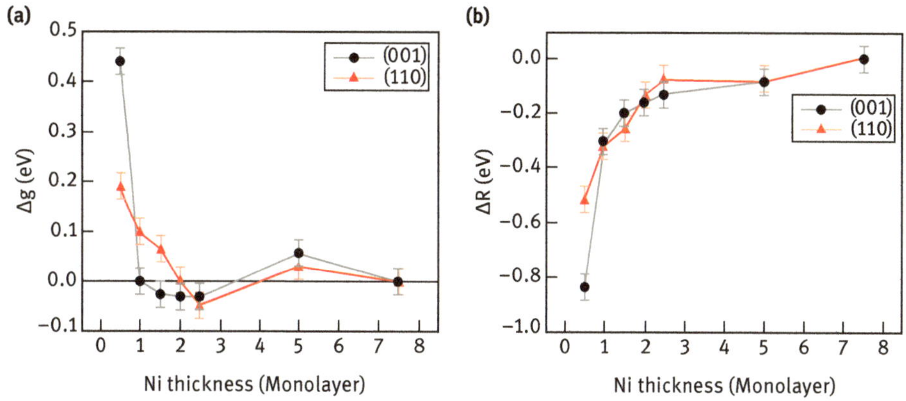

Fig. 3.7: The initial (a) and final (b) state effects contribution to the total shift as a function of Ni coverage for both TiO_2(001) (solid squares) and TiO_2(110) (solid triangles) surfaces. Reproduced from [7].

It was thus possible to evidence charge transfer between the Ni clusters and the TiO_2 substrates. It explains partly initial state effects, while screening dominates the final state shifts in larger clusters.

A higher (and in some cases different) reactivity under UHV/X-ray irradiation conditions of an XPS experiment can be exhibited by nanoparticles in comparison to bulk systems, due to a more extended surface. This aspect must be kept in mind both for properly handling samples and for interpreting results.

For example, in the study [3] already cited devoted to iron-metal core oxide-shell nanoparticles it was observed that also a short exposure to air was able to deeply oxidize the core (Fig. 3.8).

Finally, 3 nm ceria nanoparticles (Fig. 3.9) showed very pronounced reduction to Ce^{3+} under X-ray irradiation, where a very small effect was measured on thin-film form (Fig. 3.9).

3.3 Conclusions

Application of XPS to nanostructured systems can produce novel pieces of information. Both morphological and chemical aspects can be investigated using the nanometer information depth of the technique and its peculiar chemical shift effect. However, great efforts are necessary for this field to pass from the present specialist realm into routine application of the technique:

a. the application to more numerous and diverse systems needs to establish a database;
b. deeper knowledge of phenomena responsible for peculiar XPS effects in the nanometer domain is required;

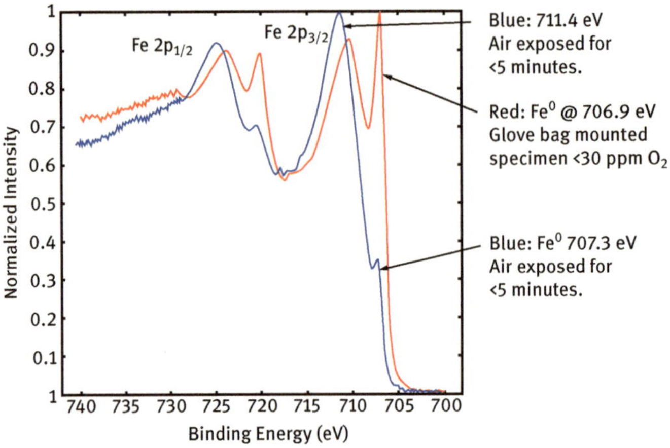

Fig. 3.8: Fe 2p photoelectron spectra from iron metal-core oxide-shell nanoparticles showing the effect of different handling methods. The first spectrum was measured after removal from storage in a glove box and transferred to an XPS system without exposure to air. The second blue spectrum shows the reduction of metallic iron ($\simeq$707 eV) after a few minutes of exposure to air. Reproduced from [7].

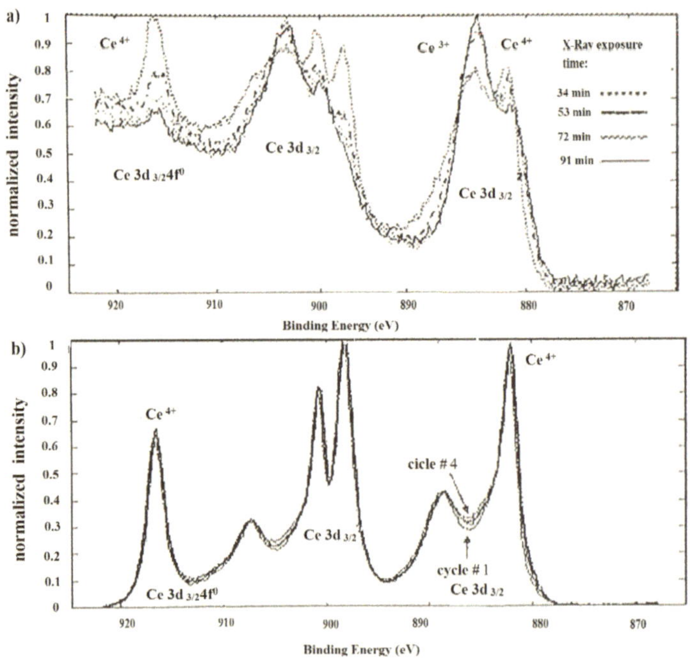

Fig. 3.9: (a) 3d photoelectron spectra of 3 nm ceria particles as examined in a Phi Quantum 2000. Note that the amount of Ce^{3+} increases with x-ray exposure; (b) very little Ce^{3+} is found in the thin film, although some very small amount of damage may occur with prolonged x-ray exposure. Adapted from [9].

c. deeper knowledge will permit to develop robust models which could also be applied by untrained researchers to extract relevant information from XPS data.

Acknowledgments

Support by the Project PON 254/Ric. Potenziamento del CENTRO RICERCHE PER LA "SALUTE DELL'UOMO E DELL'AMBIENTE" (Cod.PONa3_00334) is gratefully acknowledged

Bibliography

[1] Finster J, Lorenz P, Meisel A. Quantitative ESCA surface analysis applied to catalysts: Investigation of concentration gradients. Surface and Interface Analysis, 1979; 1: 179–184.
[2] Srivastava S. ESCA studies of supported catalysts. Applied Spectroscopy Reviews, 1988; 24: 81–97.
[3] Bear DR , Engelhard MH. XPS analysis of nanostructured materials and biological surfaces. Journal of Electron Spectroscopy and related Phenomena, 2010; 178–179: 415–432.
[4] Yang D-Q, Meunier M, Sacher E. The estimation of the average dimensions of deposited clusters from XPS emission intensity ratios. Applied Surface Science, 2001; 173: 134–139.
[5] Seah MP, Spencer SJ, Bensebaa F et al. Critical review of the current status of thickness measurements for ultrathin SiO_2 on Si Part V: Results of a CCQM pilot study. Surface and Interface Analysis, 2004; 36: 1269–1303.
[6] Hajati S, Zaporojtchenko V, Faupel F, Tougaard S. Characterization of Au nano-cluster formation on and diffusion in polystyrene using XPS peak shape analysis. Surface Science, 2007; 601: 3261–3267.
[7] Tao JG, Pan JS, Huan CHA, Zhang Z, Chai JW, Wang SJ. Origin of XPS binding energy shifts in Ni clusters and atoms on rutile TiO_2 surfaces. Surface Science, 2008; 602: 2769–2773.
[8] Richter B, Kuhlenbeck H, Freund HJ, Bagus PS. Cluster core-level binding-energy shifts: the role of lattice strain, Physical Review Letters, 2004; 93: 026805.
[9] Baer DR, et al. Challenges in applying surface analysis methods to nanoparticles and nanostructured materials. Journal of Surface Analysis, 2005; 12: 101–108.

Angela Altomare, Corrado Cuocci, Anna Moliterni, and
Rosanna Rizzi

4 Single crystal and powder XRD techniques: An overview

Introduction

X-ray diffraction (XRD) technique is nowadays widely used for solving crystal structures. In the case of single crystal availability, small and medium size structures (up to 300 non-hydrogen atoms in the asymmetric unit), as well as small proteins (provided high-quality data are available, to about 1.1 Å resolution), can be quite routinely solved. Unfortunately, crystalline materials cannot always be prepared or are not available as single crystals of dimensions and/or quality suitable to XRD technique. In these cases, powder XRD technique (PXRD) represents the preferred choice for structure characterization. In spite of unavoidable difficulties related to PXRD with respect to the single crystal case (mainly due to reflection overlap, wrong background definition and preferred orientation effects), in the last 20 years, the progress in experimental devices and the development of new and effective methodologies have led to a great and growing application of PXRD.

This chapter aims at providing a general overview of XRD technique by both single crystal and powder. In particular, the basic principles of the diffraction theory, some fundamental tools in crystallography and the main steps of the single crystal solution process by direct methods are given in Section 4.1. Section 4.2, devoted to PXRD, deals mainly with theoretical and computing approaches and problems concerning the full pathway of the solution process by PXRD.

The efficiency of the XRD technique is demonstrated by examples of successful crystal structure solutions.

4.1 The single crystal XRD technique

Introduction

The theory of X-ray diffraction provides the mathematical tools for obtaining the crystal structure from diffraction experiment. The electron density is the inverse Fourier transform of the amplitudes of the scattered waves [1]. If we knew the amplitudes, their moduli and phases, the attainment of the solution would be easy: the peaks in the electron density map provide the location of atoms in the crystal cell (the chemical species of the atom is deduced from the peak intensity). Unfortunately, the diffraction experiment gives only the intensity of the scattered amplitude but not the phase. Under the assumption of the positivity and atomicity of the electron density, direct methods [2] are able to derive the phases directly from the experimental intensities, through probabilistic mathematical relationships. In the case of single crystal (mm dimension) data of small and medium size molecules, the experimental diffraction intensities are quite accurately measured and associated to each diffraction effect. Several computing programs are available: starting from the chemical formula and the intensities, they provide the structure solution in a fast and automatic way.

Since the theory of diffraction is the same for all kinds of radiation, for the sake of brevity we will focus our attention on X-rays only; a similar approach can be adopted if a different source is considered (*i.e.*, neutrons or electrons). The theory of diffraction is described in Section 4.1.1 in accordance to the kinematical approximation, based on the assumption that the photons of the primary X-ray beam striking the matter scatter once and the interaction of the scattered radiation with the matter is negligible. If absorption phenomena are also neglected, two main and different effects can occur: coherent scattering (Thomson effect) and incoherent scattering (Compton effect), both of them will be briefly described in Section 4.1.1 where also a description of the radiation-matter interaction in the case of an extended body is given. Section 4.1.2 is devoted to some very basic concepts in crystallography and to X-ray diffraction by crystal. Section 4.1.3 regards the phase problem and crystal structure solution by direct methods (included an example of application).

4.1.1 Basics of the radiation-matter interaction

Thomson scattering

Let us consider an incident plane monochromatic electromagnetic wave striking a free particle having electric charge e and mass m, located at the origin O of a three-dimensional reference system. The particle becomes a source of electromagnetic radiation (scattered radiation), in accordance to the classical theory of electromagnetism. Thomson demonstrated that, in the case of a non-polarized primary beam, the intensity of the scattered radiation at an observation point, whose distance from the origin O is r, is given by

$$\mathrm{I}_{\mathrm{eTh}} = \mathrm{I}_{\mathrm{i}} \cdot \frac{e^4}{m^2 r^2 c^4} \cdot \frac{1 + \cos^2 2\theta}{2}, \tag{4.1}$$

where I_i is the intensity of the incident radiation, c is the speed of the light and 2θ is the angle between the direction of observation and that of the incident wave propagation.

From (4.1) it can be easily deduced that the scattered radiation is maximum along the direction of the primary beam $2\theta = 0$, and is minimum in the direction perpendicular to it ($2\theta = \pi/2$); in addition, the scattering is coherent (*i.e.*, a well-defined phase relationship between the incident and scattered waves exists) and elastic (no energy loss during the process). The presence of m at the denominator of (4.1) makes negligible the contribution of protons with respect to that of electrons. Consequently, the Thomson effect gives information mainly on the electron density; for that reason e in (4.1) traditionally stands for the electron charge.

Compton scattering

In accordance to the corpuscular nature of waves, the Compton scattering consists of a collision between a photon and a free electron characterized by an energy transfer from the incident photon to the electron. Contrary to the Thomson effect, no phase relationship exists between the incident and scattered beam, consequently the scattering is incoherent. The energy loss of the primary photon is responsible for the wavelength difference between the scattered and the incident wave, given (in Å) by $\Delta\lambda = 0.024 \cdot (1 - \cos 2\theta)$. Therefore $\Delta\lambda$ is independent of the wavelength of the incident X-ray beam and attains its maximum and minimum values at $2\theta = \pi$ (backscattering) and $2\theta = 0$, respectively. In the case of crystal, the Compton effect contributes to the background radiation, consequently it will be neglected hereafter.

Interaction of radiation – extended body

If the matter is an extended body, interference is the main phenomenon that governs the radiation-matter interaction. For the sake of simplicity, let us consider two charged particles in O and O', interacting with an incident X-ray beam (Fig. 4.1). The phase difference between the waves scattered by the electrons in O′ and O in the direction of $\mathbf{s}$ ($\mathbf{s}_0$ is the direction of the incident wave) is given by

$$\delta = \frac{2\pi}{\lambda}(\mathbf{s}_0 - \mathbf{s}) \cdot \mathbf{r}. \tag{4.2}$$

If the vector $\mathbf{r}^* = \frac{(\mathbf{s}_0 - \mathbf{s})}{\lambda}$, called the reciprocal vector, with modulus $|\mathbf{r}^*| = \frac{2\sin\theta}{\lambda}$ (Fig. 4.1), is introduced in (4.2), then $\delta = 2\pi\mathbf{r}^* \cdot \mathbf{r}$. If A_O and $A_{O'} \exp(2\pi i \mathbf{r}^* \cdot \mathbf{r})$ are the amplitudes of the wave scattered in O and O′, respectively, the total amplitude of the scattered wave is given by:

$$F(\mathbf{r}^*) = A_O + A_{O'} \exp(2\pi i \mathbf{r}^* \cdot \mathbf{r}). \tag{4.3}$$

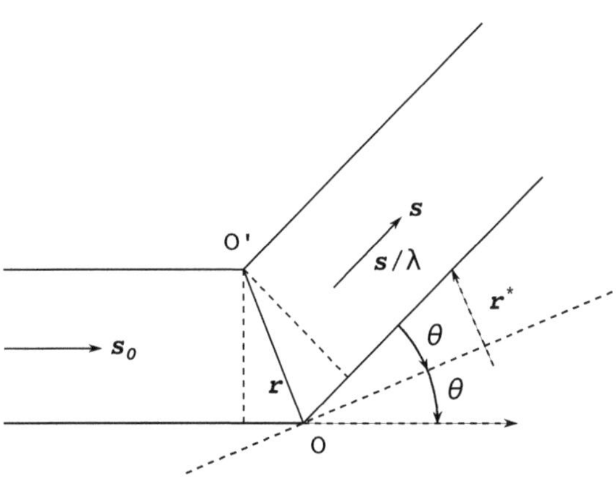

Fig. 4.1: Two scattering centers in O and O′. The primary and diffuse wave propagation directions are defined by the unit vectors $\mathbf{s}_0$ and $\mathbf{s}$, respectively.

In the case of N point scatterers, (4.3) is generalized as follows:

$$F(\boldsymbol{r}^*) = \sum_{\mathrm{j}=1}^{\mathrm{N}} f_{\mathrm{j}} \exp(2\pi \mathrm{i} \boldsymbol{r}^* \cdot r_j), \qquad (4.4)$$

where f_{j} and $\boldsymbol{r}_{\mathrm{j}}$ are the scattering factor and the position vector of the j-th scatterer, respectively. The scattering factor is defined according to $f_{\mathrm{j}}^2 = \frac{\mathrm{I_j}}{\mathrm{I_{eTh}}}$, where $\mathrm{I_j}$ is the intensity of the wave scattered by the j-th scatterer.

If the scatterers constitute a continuous system, (4.4) becomes:

$$F(\boldsymbol{r}^*) = \int_{\mathrm{S}} \rho(\boldsymbol{r}) \exp(2\pi \mathrm{i} \boldsymbol{r}^* \cdot \boldsymbol{r}) \mathrm{d}\boldsymbol{r} = T[\rho(\boldsymbol{r})], \qquad (4.5)$$

where S is the space of $\boldsymbol{r}$ vector ($\boldsymbol{r}$ is a vector in direct space). From (4.5) it can be deduced that the amplitude of the scattered wave $F(\boldsymbol{r}^*)$ is the Fourier transform (T) of the density of the elementary scatterers (*i.e.*, the electron density $\rho(\boldsymbol{r})$). Vice versa, by applying the theory of Fourier transform, the electron density $\rho(\boldsymbol{r})$ can be derived from $F(\boldsymbol{r}^*)$ by the inverse Fourier transform, as follows:

$$\rho(\boldsymbol{r}) = \int_{\mathrm{S}^*} F(\boldsymbol{r}^*) \exp(-2\pi \mathrm{i} \boldsymbol{r}^* \cdot \boldsymbol{r}) \mathrm{d}\boldsymbol{r}^* = T^{-1}[F(\boldsymbol{r}^*)], \qquad (4.6)$$

where S^* is the space of the $\boldsymbol{r}^*$ vector ($\boldsymbol{r}^*$ is a vector in reciprocal space). Equations (4.5) and (4.6) suggest that there is a biunique relationship between $\rho(\boldsymbol{r})$ and $F(\boldsymbol{r}^*)$: if $\rho(\boldsymbol{r})$ is known it is possible to derive $F(\boldsymbol{r}^*)$ and vice versa. Equations (4.5) and (4.6) are quite general, consequently they are still valid if $\rho(\boldsymbol{r})$ represents the electron density of an electron, of an atom, of a molecule, of the unit cell and of the full crystal. In the case of one electron in an atom, the electron density is described by the distribution function $\rho_{\mathrm{e}}(\boldsymbol{r}) = |\psi(\boldsymbol{r})|^2$, where $\psi(\boldsymbol{r})$ is the wave function satisfying the Schrödinger equation. The electron scattering factor $\mathrm{f_e}$ is given by $T[\rho_{\mathrm{e}}(\boldsymbol{r})]$. If the interaction with X-rays involves an atom with N electrons, the atomic scattering factor is given by the contribution of all the electrons in the atom:

$$f_{\mathrm{a}}(\boldsymbol{r}^*) = \sum_{\mathrm{j}=1}^{\mathrm{N}} f_{\mathrm{ej}}. \qquad (4.7)$$

The dependence of the scattering factor *vs.* $\sin\theta/\lambda$ in cases of C, Mg^{++}, and Cl atoms is shown in Figure 4.2; it can be noted that the maximum value of f_{a} is equal to the atomic number when $\sin\theta/\lambda = 0$ and decreases with increasing $\sin\theta/\lambda$.

If the matter consists of a molecule of N atoms, the amplitude of the scattered wave takes into account each atom in the molecule, as follows:

$$F_{\mathrm{M}}(\boldsymbol{r}^*) = \sum_{\mathrm{j}=1}^{\mathrm{N}} f_{\mathrm{j}}(\boldsymbol{r}^*) \exp(2\pi \mathrm{i} \boldsymbol{r}^* \cdot \boldsymbol{r}_{\mathrm{j}}), \qquad (4.8)$$

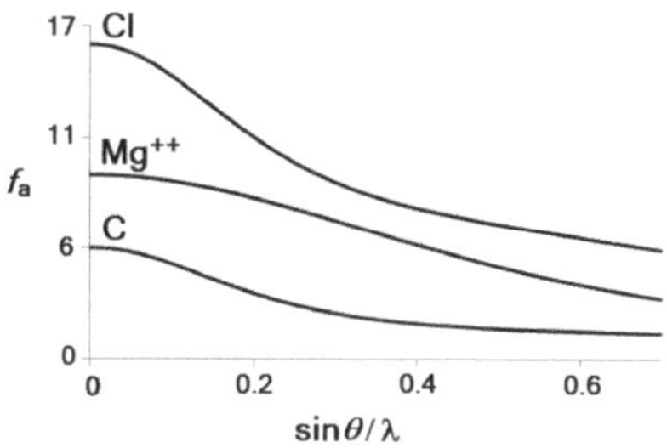

Fig. 4.2: Scattering factors *vs.* $\sin\theta/\lambda$ for C, Mg^{++}, and Cl atoms.

where $\boldsymbol{r}_j$ and $f_j(\boldsymbol{r}^*)$ are the position vector and the scattering factor of the j-th atom, respectively.

If the atomic thermal motion is taken into account, the atomic scattering factor is given by:

$$f_a(\boldsymbol{r}^*) = {}^0f_a(\boldsymbol{r}^*)\exp\left(-\mathrm{B}\frac{\sin^2\theta}{\lambda^2}\right), \tag{4.9}$$

where ${}^0f_a(\boldsymbol{r}^*)$ is the scattering factor in cases of an atom at rest, B is the isotropic thermal factor, *i.e.*, it is assumed that the atom vibrates equally in all the directions (the thermal motion has a spherical symmetry); $\mathrm{B} = 8\pi^2 u$ and u is the atomic square mean shift of the atom from the equilibrium position. In real cases, the isotropic hypothesis is not satisfied and it is preferred that B, instead of a constant, is described by six parameters, defining the lengths of the axes and the orientation of a thermal ellipsoid, centered at the mean atomic position.

4.1.2 Basics of crystallography and X-ray diffraction by crystal

The main aim of this section is to describe the radiation-crystal interaction. First some basic concepts in crystallography are given.

Crystalline state

The crystalline state is characterized by a regular repetition of a motif (*i.e.*, an atom, a molecule ...) in three-dimensional space. If each motif is substituted by a point (*e.g.*, placed at its center of gravity), the crystal can be described by a collection of periodically ordered points (*i.e.*, the lattice); the lattice points are called nodes.

If a two-dimensional crystal is considered as an example (Fig. 4.3a), two non-collinear lattice translations (*i.e.*, the vectors $\boldsymbol{a}$ and $\boldsymbol{b}$) with a common origin identify a unit cell, that is called *primitive* when all the nodes constitute its vertices (Fig. 4.3b). Each of the four vertices in a two-dimensional primitive cell belongs to it for $\frac{1}{4}$, consequently it contains only one node, on the contrary, *multiple or centered* cells contain more than one node (Fig. 4.3c).

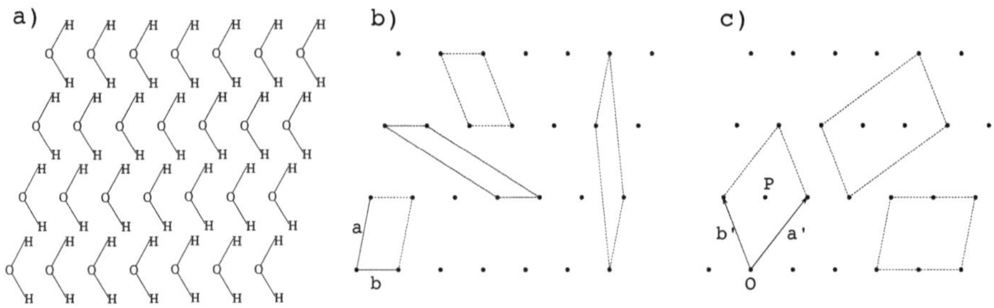

Fig. 4.3: An example of two-dimensional crystal (a) and corresponding lattice with some primitive (b) and multiple cells (c).

In the case of three-dimensional space, the cell is identified by three non-collinear vectors $\{\boldsymbol{a}, \boldsymbol{b}, \boldsymbol{c}\}$; the angles between the pairs of base vectors $(\boldsymbol{b}, \boldsymbol{c})$, $(\boldsymbol{c}, \boldsymbol{a})$, and $(\boldsymbol{a}, \boldsymbol{b})$ are α, β, and γ, respectively. Each lattice point can be uniquely defined *via* a linear combination of the base vectors, *i.e.*, by $\boldsymbol{r}_{u,v,w} = u\boldsymbol{a} + v\boldsymbol{b} + w\boldsymbol{c}$, with u, v, and w integer numbers. A three-dimensional crystal can be described by the regular repetition of the unit cell. Crystals defined by similar cells (*i.e.*, with same conditions on the ratios $a : b : c$ and on α, β, and γ angles) can be grouped into seven different crystal systems (Tab. 4.1).

Tab. 4.1: The seven crystal systems.

Crystal system	Cell axes relationships	Cell angles relationships
Triclinic	$a \neq b \neq c$	$\alpha \neq \beta \neq \gamma$
Monoclinic	$a \neq b \neq c$	$\alpha = \gamma = 90°, \beta \neq 90°$
Orthorhombic	$a \neq b \neq c$	$\alpha = \beta = \gamma = 90°$
Tetragonal	$a = b \neq c$	$\alpha = \beta = \gamma = 90°$
Trigonal	$a = b = c$	$\alpha = \beta = \gamma \neq 90°$
Hexagonal	$a = b \neq c$	$\alpha = \beta = 90°, \gamma = 120°$
Cubic	$a = b = c$	$\alpha = \beta = \gamma = 90°$

Symmetry in direct space

The crystal can show, in addition to the lattice periodicity, other types of symmetries. A symmetry operation leaves unchanged all the properties of the space after its application. Examples of symmetry operations are

- translation (Fig. 4.4a);
- rotation around an axis of n order, in this case the symmetry operation consists of a rotation of $2\pi/n$ around the axis (Fig. 4.4b);
- rototranslation, *i.e.*, the product of a rotation around an axis by a translation of **T** along it (Fig. 4.4c);
- inversion with respect to a point (Fig. 4.4d);

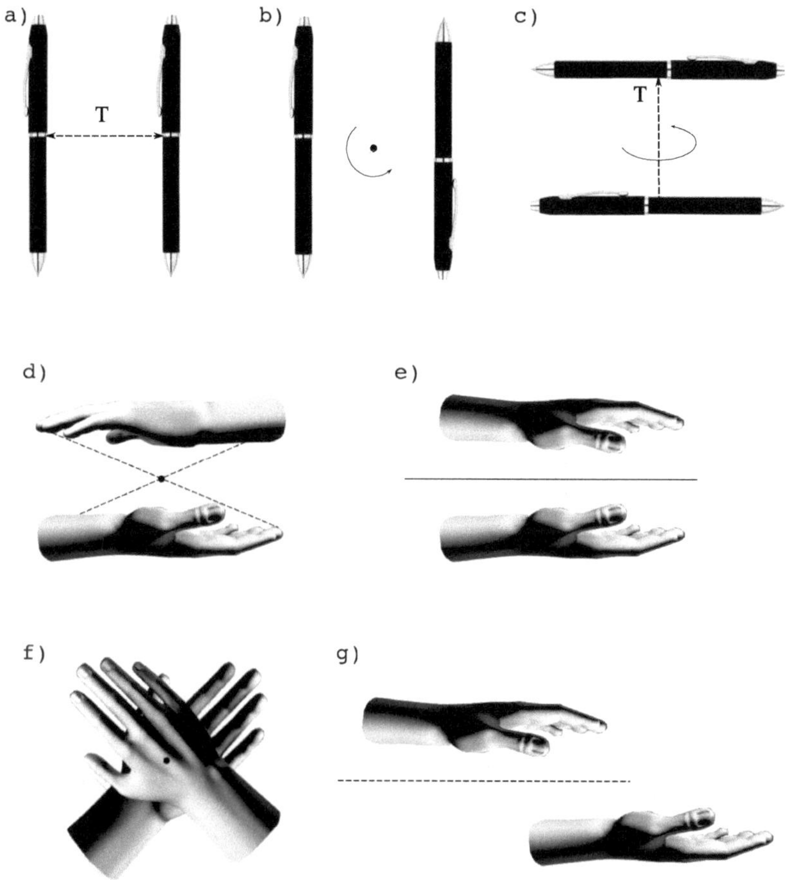

Fig. 4.4: Examples of symmetry operations: translation (a), rotation (b), rototranslation (c), inversion with respect to a point (d), reflection with respect to a plane (e), rotoinversion (anticlockwise rotation) or rotoreflection (clockwise rotation) (f), glide plane (g).

- reflection by a plane (Fig. 4.4e);
- rotoinversion, *i.e.*, the product of a rotation around an axis by an inversion with respect to a point belonging to the axis (Fig. 4.4f);
- rotoreflection, *i.e.*, the product of a rotation around an axis by a reflection by a plane normal to the axis (Fig. 4.4f);
- glide plane, *i.e.*, the product of a reflection by a plane by a translation parallel to the plane (Fig. 4.4g).

In order to satisfy the lattice periodicity, not all the rotations, rototranslations and glide planes are allowed. For example, in the case of rotations the allowed symmetry operations correspond to $n = 1, 2, 3, 4, 6$ and the forbidden one to $n = 5$.

The set of geometrical symmetry operations that take a three-dimensional periodic object (*i.e.*, a crystal) into itself constitutes a crystallographic space group. If all the combinations of the symmetry operations together with the possible cells (primitive or centered) are considered, the total number of crystallographic space groups is 230. When the symmetry operations are taken into account, instead of the unit cell, a portion of unit cell, the smallest part, called asymmetric unit can be considered. The whole cell content is generated by applying all the symmetry operations to the asymmetric unit content.

An additional concept, basic for the diffraction by crystal, concerns the Miller indices $\mathbf{h} = (hkl)$, also called reflection, that are three integer numbers identifying a family of lattice planes in the direct space. They divide the axis $\boldsymbol{a}$ into h parts, $\boldsymbol{b}$ into k parts, and $\boldsymbol{c}$ into l parts.

Reciprocal space

The reciprocal lattice represents a fundamental tool for describing the diffraction by crystal. It is defined by three base vectors $\{\boldsymbol{a}^*, \boldsymbol{b}^*, \boldsymbol{c}^*\}$, related to the bases $\{\boldsymbol{a}, \boldsymbol{b}, \boldsymbol{c}\}$ of the direct (or real) lattice by the following metric relationships:

$$\begin{aligned} \boldsymbol{a}^* \cdot \boldsymbol{a} = \boldsymbol{b}^* \cdot \boldsymbol{b} = \boldsymbol{c}^* \cdot \boldsymbol{c} = 1, \\ \boldsymbol{a}^* \cdot \boldsymbol{b} = \boldsymbol{a}^* \cdot \boldsymbol{c} = \boldsymbol{b}^* \cdot \boldsymbol{a} = \boldsymbol{b}^* \cdot \boldsymbol{c} = \boldsymbol{c}^* \cdot \boldsymbol{a} = \boldsymbol{c}^* \cdot \boldsymbol{b} = 0. \end{aligned} \tag{4.10}$$

The relationships (4.10) suggest that $\boldsymbol{a}^* \perp \boldsymbol{b}$ and $\boldsymbol{c}$, $b^* \perp \boldsymbol{a}$ and $\boldsymbol{c}$, and $c^* \perp \boldsymbol{a}$ and $\boldsymbol{b}$.

A generic lattice vector $\boldsymbol{r}^*_{\mathbf{h}}$ in the reciprocal space, in terms of the base vectors $\{\boldsymbol{a}^*, \boldsymbol{b}^*, \boldsymbol{c}^*\}$, is given by

$$\boldsymbol{r}^*_{\mathbf{h}} = h\boldsymbol{a}^* + k\boldsymbol{b}^* + l\boldsymbol{c}^*. \tag{4.11}$$

It can be easily demonstrated that

- due to (4.10), $\boldsymbol{r}^*_{\mathbf{h}}$ is perpendicular to the family of lattice planes (hkl) in direct space;
- its modulus $|\boldsymbol{r}^*_{\mathbf{h}}| = 1/d_{\mathbf{h}}$; $d_{\mathbf{h}}$ is the interplanar distance of the planes (hkl).

Diffraction by crystal

The two cases corresponding to the infinite and the finite crystal follow.

Infinite crystal

The periodical repetition in an infinite crystal can be described by the position vector $\boldsymbol{r}_{u,v,w}$ of the lattice point. Consequently, an infinite crystal can be mathematically represented *via* an infinite summation of δ Dirac functions, providing the following lattice function:

$$L(\boldsymbol{r}) = \sum_{u=-\infty}^{+\infty} \sum_{v=-\infty}^{+\infty} \sum_{w=-\infty}^{+\infty} \delta(\boldsymbol{r} - \boldsymbol{r}_{u,v,w}).$$

It can be demonstrated [1] that the electron density of an infinite crystal is the convolution of the electron density in the unit cell $\rho_M(\boldsymbol{r})$ with the lattice function $L(\boldsymbol{r})$:

$$\rho \boldsymbol{r}_{\infty}(\boldsymbol{r}) = \rho_M(\boldsymbol{r}) * L(\boldsymbol{r}). \tag{4.12}$$

Equation (4.12) is represented by an example in the two-dimensional case (Fig. 4.5).

Fig. 4.5: Representation of a two-dimensional crystal (c) as convolution of electron density in the unit cell $\rho_M(\mathbf{r})$ (a) with a lattice (b).

If the Fourier transform of (4.12) is calculated, and the properties of convolution are taken into account, the following results are obtained:

$$\begin{aligned} F_{\infty}(\boldsymbol{r}^*) = T[\rho_{\infty}(\boldsymbol{r})] &= \frac{1}{V} F_M(\boldsymbol{r}^*) \sum_{h,k,l=-\infty}^{+\infty} \delta(\boldsymbol{r}^* - \boldsymbol{r}^*_{\mathbf{h}}) \\ &= \frac{1}{V} F_M(\mathbf{h}) \sum_{h,k,l=-\infty}^{+\infty} \delta(\boldsymbol{r}^* - \boldsymbol{r}^*_{\mathbf{h}}), \end{aligned} \tag{4.13}$$

where V is the volume of the unit cell associated to the direct lattice, $F_M(\boldsymbol{r}^*)$ is given by (4.8), and $\boldsymbol{r}^*_{\mathbf{h}}$ is the lattice vector of the reciprocal lattice (Eq. (4.11)).

In accordance to Section 4.1.1, Eq. (4.13) represents the total amplitude of the wave scattered by the crystal and suggests that, due to the δ function properties, in the case of a crystal the diffraction effects occur only when

$$\boldsymbol{r}^* = \boldsymbol{r}^*_{\mathbf{h}}, \tag{4.14}$$

i.e., when $\boldsymbol{r}^*$ is the position vector of a reciprocal lattice point, emphasizing the important role of the reciprocal lattice in the diffraction theory.

Finite crystal
A finite crystal may be represented by the following product:

$$\rho_{cr}(\boldsymbol{r}) = \rho_{\infty}(\boldsymbol{r}) \cdot \Phi(\boldsymbol{r}), \tag{4.15}$$

where $\Phi(\boldsymbol{r})$ is the form function, and it is equal to 1 inside the crystal, to 0 otherwise. The total amplitude of the diffracted wave is calculated by the Fourier transform of (4.15) (Section 4.1.1) and by taking into account the properties of convolution, as follows:

$$F(\boldsymbol{r}*) = T[\rho_{cr}(\boldsymbol{r})] = T[\rho_{\infty}(\boldsymbol{r}) \cdot \Phi(\boldsymbol{r})]$$
$$= T[\rho_{\infty}(\boldsymbol{r})] * T[\Phi(\boldsymbol{r})] = \frac{1}{V} F_{M}(\mathbf{h}) \sum_{h,k,l=-\infty}^{+\infty} D(\boldsymbol{r}^{*} - \boldsymbol{r}^{*}_{\mathbf{h}}), \quad (4.16)$$

where $D(\boldsymbol{r}^*)$ is the Fourier transform of the form function $\Phi(\boldsymbol{r})$.

Equations (4.13) and (4.16) suggest that going from an infinite to a finite crystal, the δ function (corresponding to each reciprocal lattice node) is replaced by the distribution function D which is different from 0 in a domain whose shape and dimension depend on the crystal shape and dimensions. In particular, from the properties of the Fourier transform it can be derived that the width of a diffraction principle maximum in a certain direction is inversely proportional to the dimension of the crystal in that direction. Due to their small dimensions, this effect is critical in the case of nanocrystals.

The term $F_M(\mathbf{h})$ in Eqs. (4.13) and (4.16) is called the structure factor. For the sake of brevity it will be denoted, from now on, as $F_{\mathbf{h}}$. It is given by

$$F_{\mathbf{h}} = \sum_{j=1}^{N} f_j \exp(2\pi i \boldsymbol{r}^{*}_{\mathbf{h}} \cdot \boldsymbol{r}_j) = \sum_{j=1}^{N} f_j \exp(2\pi i \mathbf{h} \cdot \mathbf{X}_j) = A_{\mathbf{h}} + iB_{\mathbf{h}}, \quad (4.17)$$

where $A_{\mathbf{h}} = \sum_{j=1}^{N} f_j \cos(2\pi\mathbf{h} \cdot \mathbf{X}_j)$, $B_{\mathbf{h}} = \sum_{j=1}^{N} f_j \sin(2\pi\mathbf{h} \cdot \mathbf{X}_j)$, N is the number of atoms in the unit cell, $\boldsymbol{r}_j$ the positional vector of the j-th atom, and $\mathbf{X}_j$ represents the transpose matrix of the components of $\boldsymbol{r}_j$ in a direct coordinate system. Equation (4.17) suggests that $F_{\mathbf{h}}$ is a complex quantity, consequently it can be expressed in terms of modulus and phase, as follows:

$$F_{\mathbf{h}} = |F_{\mathbf{h}}| \exp(i\varphi_{\mathbf{h}}), \quad \text{with} \quad \varphi_{\mathbf{h}} = \operatorname{atan}(B_{\mathbf{h}}/A_{\mathbf{h}}),$$

and therefore, in order to determine the electron density $\rho(\boldsymbol{r})$ we have to know the structure factors in moduli and phases.

The diffraction by crystal is schematized in Figure 4.6: when X-rays strike the crystal, diffraction effects, corresponding to the Fourier transform of the crystal electron density, occur. The electron density of the crystal can be determined by the inverse Fourier transform that, applied to the amplitude of the wave scattered by crystal, is able to reconstruct the image of the scatterers, like a mathematical lens.

The Bragg law

The diffraction by crystal has been described in a simple and qualitative way by Bragg in terms of reflection of the incident wave by the atoms lying on the lattice planes with Miller indices $\mathbf{h} = (hkl)$ and interplanar distance $d_{\mathbf{h}}$. In order to have constructive interference between the beams diffracted by atoms in D and B (Fig. 4.7), the difference

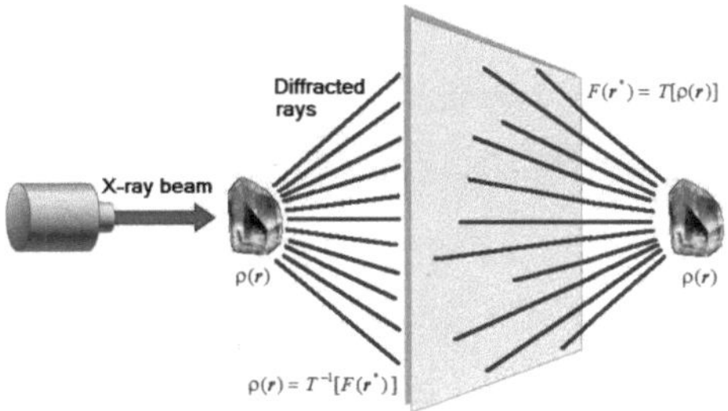

Fig. 4.6: epresentation of the diffraction by crystal.

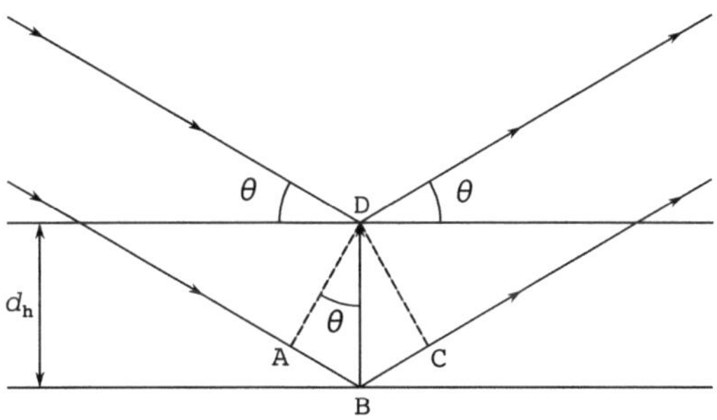

Fig. 4.7: Bragg representation of X-ray diffraction by lattice planes with interplanar distance d_h

of optical path (AB + BC) should be a multiple of the wavelength λ, *i.e.*,

$$2d_h \sin\theta = n\lambda, \quad n = 1, 2 \ldots . \tag{4.18}$$

From (4.18) the following relation can be easily derived:

$$\frac{2\sin\theta}{\lambda} = \frac{1}{d_h},$$

which coincides with the condition (4.14), when the vector moduli only are taken into account.

The Ewald sphere

An elegant description of the diffraction by crystal has been proposed by Ewald by introducing a sphere of radius 1/λ (the so called Ewald sphere). Let us consider O to be the origin of the reciprocal space (Fig. 4.8), IO the direction of the incident X-ray beam, and P a reciprocal lattice point whose position vector $\boldsymbol{r}_h^*$ is perpendicular to the family of planes (*hkl*). Ewald demonstrated that diffraction occurs when the point P belongs to the Ewald sphere, in this case the family of planes (*hkl*) involved in the diffraction lies parallel to IP and the following relationship can be derived (Fig. 4.8):

$$OP = |\mathbf{r}_{\mathbf{h}}^{*}| = 1/d_{\mathbf{h}} = \mathrm{IO}\sin\theta = 2\sin\theta/\lambda. \tag{4.19}$$

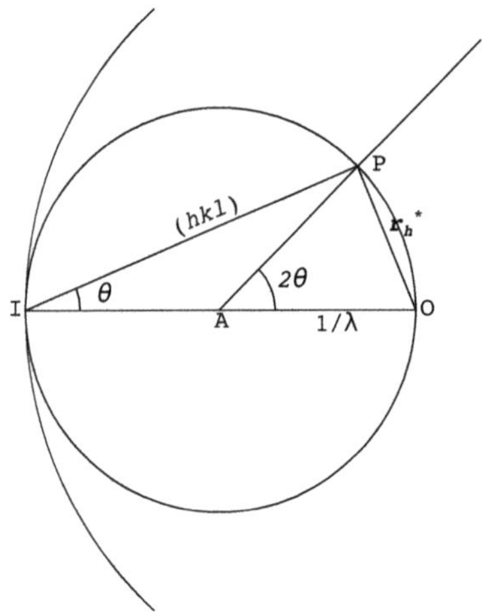

Fig. 4.8: X-ray diffraction by crystal described *via* the Ewald sphere.

Equation (4.19) states that when a reciprocal lattice point is on the Ewald sphere, the Bragg condition is satisfied. From (4.19), since $|\sin\theta| \leq 1$, the following restriction arises: $|\mathbf{r}_{\mathbf{h}}^{*}| \leq 2/\lambda$ and, consequently, $d_{\mathbf{h}} \geq \lambda/2$ (*i.e.*, no diffraction effects occur if $d_{\mathbf{h}} < \lambda/2$ The value $2/\lambda$ corresponds to the radius of the so called limiting sphere, centered in O: all the lattice points inside this sphere contribute to diffraction.

It can be asserted that the aim of a diffraction experiment is to bring the reciprocal lattice points onto the Ewald sphere.

It has been demonstrated that the measured intensity for each reflection is proportional to the squared structure factor modulus (*i.e.*, $I_{\mathbf{h}} \propto |F_{\mathbf{h}}|^2$).

Symmetry in the reciprocal space

The following most important properties, related to the symmetry in the reciprocal space, have to be considered.

– *The diffraction intensities show an inversion center*. Let us consider two reflections with vectors **h** and −**h**, respectively. According to (4.17), the corresponding structure factors are $F_{\mathbf{h}} = A_{\mathbf{h}} + iB_{\mathbf{h}}$ and $F_{-\mathbf{h}} = A_{\mathbf{h}} - iB_{\mathbf{h}}$, respectively, and the relationship between the corresponding phases is $\varphi_{-\mathbf{h}} = -\varphi_{\mathbf{h}}$. By taking into account that $I_{\mathbf{h}} \propto |F_{\mathbf{h}}|^2$, it can be easily demonstrated that the intensities of $F_{\mathbf{h}}$ and $F_{-\mathbf{h}}$ are the same (Friedel law); in fact,

$$I_{\mathbf{h}} \propto (A_{\mathbf{h}} + iB_{\mathbf{h}})(A_{\mathbf{h}} - iB_{\mathbf{h}}) = A_{\mathbf{h}}^2 + B_{\mathbf{h}}^2$$
$$I_{-\mathbf{h}} \propto (A_{\mathbf{h}} - iB_{\mathbf{h}})(A_{\mathbf{h}} + iB_{\mathbf{h}}) = A_{\mathbf{h}}^2 + B_{\mathbf{h}}^2.$$

- *The presence of symmetry operations is responsible for the absence of some classes of reflections.* These kinds of reflections have $|F_{\mathbf{h}}| = 0$ and are called systematically absent. If an analysis of the absent reflections is performed, it is possible to determine the space group.

4.1.3 Solving the phase problem by direct methods

A diffraction experiment provides information on the diffraction intensities and, therefore, on structure factor moduli only. As shown in Section 4.1.2, in order to determine the electron density $\rho(\boldsymbol{r})$ of a crystal, it is necessary to recover the lost information on reflection phases, *i.e.*, to solve the so called *phase problem*, a well-known problem in crystallography. Direct methods (DM) greatly contributed to the solution of the phase problem. Their role is explained by the name: to recover the phase information directly from the observed intensities, by applying mathematical relationships. DM are probabilistic methods based on three main hypotheses:
- the positivity of $\rho(\boldsymbol{r})$;
- the atomicity of $\rho(\boldsymbol{r})$ (*i.e.*, the electron density is concentrated around the nuclei);
- the random distribution of atoms in the unit cell.

The last hypothesis enables to derive the following relationship:

$$\langle |F_{\mathbf{h}}|^2 \rangle = \sum_{j=1}^{N} f_j^2.$$

Due to f_j (Eq. (4.9)), $\langle |F_{\mathbf{h}}|^2 \rangle$ depends both on the thermal factor B and on data resolution ($\sin\theta/\lambda$). In addition, the observed diffraction intensities are on a relative scale, *i.e.*, $I_{\mathbf{h}} = k|F_{\mathbf{h}}|^2$; Wilson overcame this dependence by proposing a method, the Wilson plot method [3], able to provide an estimate of the scale factor k and of the overall isotropic thermal factor and to calculate the normalized structure factor modulus $|E_{\mathbf{h}}|$ as follows:

$$|E_{\mathbf{h}}| = \left[|F_{\mathbf{h}}|^2/k \cdot \exp(-2B\sin^2\theta/\lambda^2) \sum_{j=1}^{N} {}^0f_j^2 \right]^{1/2}. \quad (4.20)$$

$|E_{\mathbf{h}}|$ has a fundamental role in DM theory. Contrary to $|F_{\mathbf{h}}|$, $|E_{\mathbf{h}}|$ is independent of $\sin\theta/\lambda$ (it can be demonstrated that $\langle |E_{\mathbf{h}}|^2 \rangle = 1$) and corresponds to an ideal structure with a point-shaped atom not thermally moved.

The Wilson statistics can also reveal the presence of a center of symmetry by comparing the theoretical and the observed distribution of $|E|$ normalized structure factor moduli. The acentric and centric theoretical probability distribution functions of $|E|$ (see their representation in Fig. 4.9) are given by

$$_1P(|E|) = 2|E|\exp(-|E|^2) \qquad \text{acentric,}$$
$$_{\bar{1}}P(|E|) = \sqrt{\frac{2}{\pi}}\exp(-|E|^2/2) \quad \text{centric.}$$

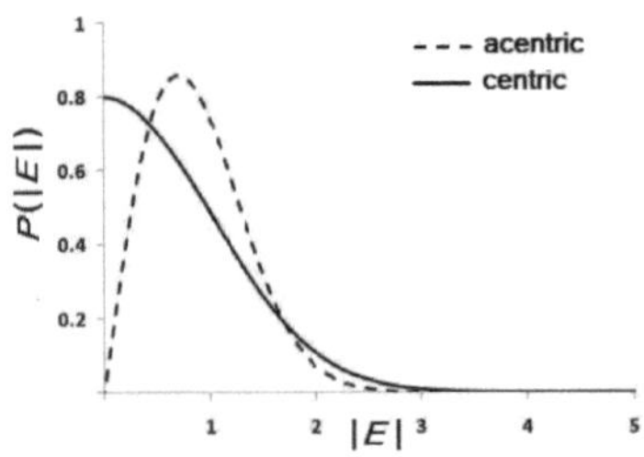

Fig. 4.9: Theoretical centric and acentric probability distribution functions of normalized structure factor moduli.

It can be observed that a diffraction experiment gives direct information on the interatomic distances, in fact

$$|F_{\mathbf{h}}|^2 = \sum_j f_j^2 + \sum_{i \neq j} f_i f_j \exp[2\pi i \mathbf{h} \cdot (\boldsymbol{r}_i - \boldsymbol{r}_j)].$$

It can be easily demonstrated that, contrary to the reflection phase, $|F_{\mathbf{h}}|^2$ is independent of the origin choice. In fact, if the origin is changed, the structure factor becomes $F'_{\mathbf{h}} = F_{\mathbf{h}} \exp(-2\pi i \mathbf{h} \cdot \boldsymbol{r}_0)$, therefore the modulus of the structure factor remains unchanged (*i.e.*, it is structure invariant) while the phase changes according to the relationship

$$\varphi'_{\mathbf{h}} = \varphi_{\mathbf{h}} - 2\pi \mathbf{h} \cdot \boldsymbol{r}_0, \tag{4.21}$$

where $\boldsymbol{r}_0$ is the position vector of the new origin with respect to the old one. Equation (4.21) suggests that the phases are not structure invariants. Let us consider a special combination of structure factors, with vectorial sum of Miller indices equal to zero; it can be demonstrated that the phase change of this combination is zero, *i.e.*, it is structure invariant. The most important structure invariant is the triplet invariant $F_{\mathbf{h}}F_{\mathbf{k}}F_{-\mathbf{h}-\mathbf{k}}$ and, in terms of normalized structure factors, $E_{\mathbf{h}}E_{\mathbf{k}}E_{-\mathbf{h}-\mathbf{k}}$, that has a crucial role in the phasing process. It can be easily verified that it does not change its phase when the origin is changed, in fact the phase change stated by (4.21) is $-2\pi[\mathbf{h} + \mathbf{k} + (-\mathbf{h} - \mathbf{k})] \cdot \boldsymbol{r}_0 = 0$. The triplet phase can be estimated by the Cochran formula [4], given by

$$\mathrm{P}(\Phi_{\mathbf{h},\mathbf{k}}) = \frac{1}{2\pi \mathrm{I}_0(\mathrm{G}_{\mathbf{h},\mathbf{k}})} \exp(\mathrm{G}_{\mathbf{h},\mathbf{k}} \cos \Phi_{\mathbf{h},\mathbf{k}}), \tag{4.22}$$

where $\Phi_{\mathbf{h},\mathbf{k}} = \varphi_{\mathbf{h}} + \varphi_{\mathbf{k}} + \varphi_{-\mathbf{h}-\mathbf{k}}$ is the phase of the triplet, $G_{\mathbf{h},\mathbf{k}}$ represents the triplet reliability $G_{\mathbf{h},\mathbf{k}} \propto |E_{\mathbf{h}}E_{\mathbf{k}}E_{-\mathbf{h}-\mathbf{k}}|$, $I_0(G_{\mathbf{h},\mathbf{k}})$ is the modified Bessel function of zero order. $P(\Phi_{\mathbf{h},\mathbf{k}})$ is a Gaussian function which assumes maximum value at $\Phi_{\mathbf{h},\mathbf{k}} \cong 0$ (*i.e.*, positive triplet), that is equivalent to the following condition:

$$\varphi_{\mathbf{h}} = -\varphi_{\mathbf{k}} - \varphi_{-\mathbf{h}-\mathbf{k}} \,. \tag{4.23}$$

The larger the $G_{\mathbf{h},\mathbf{k}}$ value, the larger the reliability of the triplet phase estimate.

The reflection **h**, with phase $\varphi_{\mathbf{h}}$, can be involved in more than one relationship like (4.23). Once the triplet phases have been estimated, if random phases are assigned to a set of selected reflections strongly involved in triplet relationships, the reflection phases can be derived *via* a chain process, by applying the tangent formula, proposed by Hauptman and Karle [5].

The tangent formula is applied in order to extend and refine the phases (convergence procedure).

At the end of the phasing process more sets of phases are available whose reliability is assessed by a suitable figure of merit (the CFOM combined figure of merit). The set corresponding to the largest CFOM is automatically selected for calculating a Fourier map (the $|E|$-map), given by:

$$\rho(\boldsymbol{r}) \cong \frac{1}{V}\sum_{\mathbf{h}} |E_{\mathbf{h}}| \exp(i\varphi_{\mathbf{h}}) \exp(-2\pi i \mathbf{h} \cdot \boldsymbol{r}). \tag{4.24}$$

The symbol $\cong$ in (4.24) has been introduced in order to emphasize that, in real cases, the summation is not extended to an infinite number of reflections but to the limited experimentally available one.

The main steps carried out by DM can be therefore so summarized:

- *normalization*, performed by the Wilson plot method. Structure factors are normalized, a subset of them is selected from among the $|E|$-largest reflections (*i.e.*, the strong reflections) and a statistical analysis of the $\{|E_{\mathbf{h}}|\}$ is performed;
- *structure invariants*. Triplets are calculated among the strong reflections and their phases evaluated;
- *phasing process*. Starting random values of phase are assigned to reflections in order to trigger the phasing process by the tangent formula. At the end of phase determination the most reliable phasing sets (*i.e.*, with the largest CFOM values) are stored;
- *E-map calculation*. The best phasing set is selected for calculating the electron density map $\rho(\boldsymbol{r})$ *via* Fourier synthesis, whose coefficients are the $|E_{\mathbf{h}}|$ values, according to (4.24). A peak search is carried out on the $|E|$-map in order to determine the peak positions to which atomic species are assigned taking into account the peak height. An automatic chemical interpretation is performed.

The structure model determined by the $|E|$-map can be incomplete, essentially due to truncation errors (*i.e.*, the number of reflections used in (4.24) is not infinite). In order to recover missing atoms and complete the structure, additional Fourier synthesis can be calculated, *e.g.*, using the phases calculated from the current model and $|F_{\mathbf{h}}|_{obs}$ or ($|F_{\mathbf{h}}|_{obs} - |F_{\mathbf{h}}|_{calc}$) coefficients (the latter is called the difference Fourier map), where $|F_{\mathbf{h}}|_{obs}$ is the measured structure factor modulus and $|F_{\mathbf{h}}|_{calc}$ is the structure factor modulus calculated by the model to be completed. The structure model is usually refined by a least-squares procedure minimizing the residual $S = \sum_{\mathbf{h}} w_{\mathbf{h}}(|F_{\mathbf{h}}|_{obs} - |F_{\mathbf{h}}|_{calc})^2$, where $w_{\mathbf{h}}$ is a suitable weight associated to each **h** reflection and taking into account the precision of the measured quantity. Atomic coordinates and thermal factors are the variables usually refined. The reliability of the structure refinement is related to the ratio "*number of observations (reflections)/number of variables to be refined*". An important agreement factor assessing the goodness of the structure model is the following residual:

$$R_F = \frac{\sum_{\mathbf{h}} ||F_{\mathbf{h}}|_{obs} - |F_{\mathbf{h}}|_{calc}|}{\sum_{h} |F_{\mathbf{h}}|_{obs}}.$$

In the case of a correct solution, very small values of R_F can be obtained (~3–6 %).

In the last part of this section a brief description of the application of DM to single crystal data is given in the case of $Ba_2Ru_2Cl_{10}O \cdot 10H_2O$ compound [6]. From the experimental diffraction pattern, the cell (monoclinic, cell parameters $a = 20.9386(18)$, $b = 8.8654(7)$, $c = 16.0560(15)$ Å, $\beta = 125.559(5)°$) as well as the diffraction intensities associated to each reflection have been determined by using the diffractometer software. An example of the collected diffraction images of $Ba_2Ru_2Cl_{10}O \cdot 10H_2O$ compound is given (Fig. 4.10) where each spot represents a reciprocal lattice node.

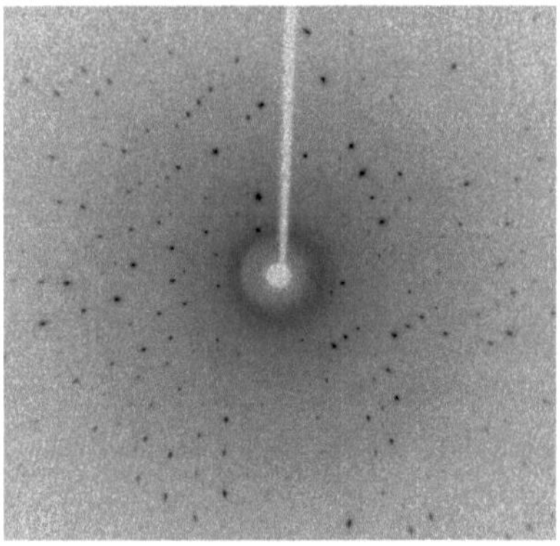

Fig. 4.10: $Ba_2Ru_2Cl_{10}O \cdot 10H_2O$ compound: an example of measured X-ray diffraction image.

Information about the symmetry of $Ba_2Ru_2Cl_{10}O \cdot 10H_2O$ compound is given (Fig. 4.11). The crystal system is monoclinic, the space group $C2/c$ means that the cell is centered, therefore not primitive, and a rotation axis of order n = 2 parallel to $\boldsymbol{b}$ is present together with a c glide normal to it. The four possible symmetry operations are also provided.

```
Crystal family                                   :     Monoclinic

Space group (centrosymmetric)                    :     C 2/c

Symmetry-operations

1)   x    ,    y    ,    z          2)   -x    ,    y    ,   -z+1/2
3)  -x    ,   -y    ,   -z          4)    x    ,   -y    ,    z+1/2
```

Fig. 4.11: $Ba_2Ru_2Cl_{10}O \cdot 10H_2O$: symmetry information.

The structure solution is carried out by DM. The total number of observed reflections is 2821; the number of strong reflections and the number of triplets are 164 and 2423, respectively. The structure model provided by DM is shown in Figure 4.12a; even if most of the atoms have been correctly positioned, the structure is incomplete, some atoms are missing and some other ones wrongly located. The complete structure model at the end of the structure refinement process is shown in the asymmetric unit (Fig. 4.12b) and in the unit cell (Fig. 4.12c), it corresponds to $R_F = 2.9\,\%$.

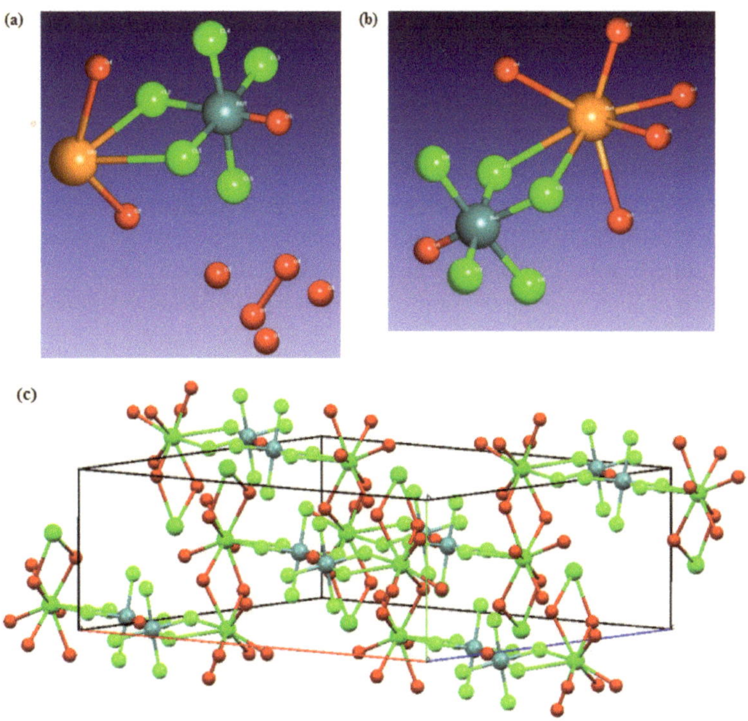

Fig. 4.12: $Ba_2Ru_2Cl_{10}O \cdot 10H_2O$ compound at the end of DM application (a), after the structure refinement in asymmetric unit (b) and in the full unit cell (c).

4.2 The powder XRD technique

Introduction

In the case of powder [7] (microcrystalline or nanocrystalline), several crystallites, randomly oriented in the space, simultaneously participate to the diffraction. As a consequence, the diffraction pattern is different from the single crystal case (one spot for each diffraction effect); it consists of a series of peaks whose intensities degrade at increasing the θ angle of diffraction (θ is the Bragg angle satisfying the diffraction condition $2 \cdot d \cdot \sin\theta = \lambda$, d interplanar distance of the crystal planes as given at Eq. (4.18)). The three-dimensional diffraction pattern of the single crystal case collapses in one dimension in the case of powder. The main problem occurring in the powder diffraction pattern is the peak overlap: peaks corresponding to different diffraction effects quite often systematically and/or casually overlap. The difficult evaluation of background and the occurrence of preferred orientation (the non-random distribution of crystallites modifies the real ratios of intensities) are additional problems whose main consequence is that the diffraction intensity cannot be accurately associated to each diffraction effect, in other words the accuracy of the experimental diffraction intensity is low. These problems (from now on *powder problems*) limit all the steps of the solution process: determination of the crystal cell, identification of the space group, intensity extraction, phasing process, electron density map calculation. Special algorithms have been developed for overcoming the overlapping limit and, in particular, for decomposing the pattern and estimating reliable intensities. In spite of great advances, both experimental and methodological, the powder solution has not yet reached the same performance as the single crystal case. *EXPO* [8] is a computing program, unique in the national and international scientific view, able to carry out automatically the full pathway of the solution process by powder. The following Sections 4.2.1–4.2.5 provide a general description of each step of the solution process from indexation to Rietveld refinement. Examples of crystal structure solution are given in Section 4.2.6.

4.2.1 Indexation

The first step in the structural solution process by PXRD is the unit cell parameters' $(a, b, c, \alpha, \beta, \gamma)$ determination, which consists of assigning the appropriate triple (hkl) of Miller indices of the crystal lattice plane to each one of the experimental diffraction peaks. In the case of powders, the collapse of the three-dimensional reciprocal space into the one-dimensional 2θ axis makes the elementary cell reconstruction more complex than in the single crystal case, because the only available information related to the observed peaks is represented by a list of experimental d_{hkl} values. They represent the observed interplanar distances and are related to the diffraction angle by the Bragg law:

$$d_{\mathbf{h}} = \lambda/2\sin\theta_{\mathbf{h}}.$$

By assuming

$$Q(hkl) = \frac{10^4}{d_{\mathbf{h}}^2}$$

the expressions of Q(*hkl*) in terms of reciprocal cell parameters and (*hkl*) indices are shown in Table 4.2 for each crystal system with

$$A_{11} = 10^4 a^{*2}, \quad A_{22} = 10^4 b^{*2}, \quad A_{33} = 10^4 c^{*2}, \quad A_{12} = 2 \cdot 10^4 a^* b^* \cos\gamma^*,$$
$$A_{13} = 2 \cdot 10^4 a^* c^* \cos\beta^*, \quad A_{23} = 2 \cdot 10^4 b^* c^* \cos\alpha^*.$$

$\boldsymbol{a}^*$, $\boldsymbol{b}^*$ and $\boldsymbol{c}^*$ are three vectors of the reciprocal lattice, related to the direct lattice ones ($\boldsymbol{a}$, $\boldsymbol{b}$, and $\boldsymbol{c}$) by

$$\boldsymbol{a} = \frac{\boldsymbol{b}^* \wedge \boldsymbol{c}^*}{V*}, \quad \boldsymbol{b} = \frac{\boldsymbol{c}^* \wedge \boldsymbol{a}^*}{V*}, \quad \boldsymbol{c} = \frac{\boldsymbol{a}^* \wedge \boldsymbol{b}^*}{V*},$$

where

$$V^* = \boldsymbol{a}^* \cdot \boldsymbol{b}^* \wedge \boldsymbol{c}^* = \frac{1}{V}$$

is the reciprocal unit cell volume.

Tab. 4.2: The expression of Q(*hkl*) depending on the crystal system.

Crystal System	Q(*hkl*)
Cubic	$(h^2 + k^2 + l^2) A_{11}$
Tetragonal	$(h^2 + k^2) A_{11} + l^2 A_{33}$
Hexagonal	$(h^2 + hk + k^2) A_{11} + l^2 A_{33}$
Orthorhombic	$h^2 A_{11} + k^2 A_{22} + l^2 A_{33}$
Monoclinic	$h^2 A_{11} + k^2 A_{22} + l^2 A_{33} + hl\, A_{13}$
Triclinic	$h^2 A_{11} + k^2 A_{22} + l^2 A_{33} + hk\, A_{12} + hl\, A_{13} + kl\, A_{23}$

In the most complicated case, the triclinic system (Tab. 4.2), powder indexing means determining six A_{ij} values, the same for each peak, and three crystallographic indices (*hkl*) different for each peak. They must satisfy the following basic indexing equation within a suitable Δ tolerance parameter:

$$Q(hkl) - \Delta < h^2 A_{11} + k^2 A_{22} + l^2 A_{33} + hk\, A_{12} + hl\, A_{13} + kl\, A_{23} < Q(hkl) + \Delta. \quad (4.25)$$

Several indexing computing programs (see [9] for an overview), using different algorithms and strategies to solve Eq. (4.25), are available.

As an example of indexation, the process followed by the *EXPO* program [8] is described, consisting of the following steps:

- a peak search procedure is executed on the PXRD profile to calculate a list of experimental interplanar distances $\{d_{\mathbf{h}}\}$. The filter method, followed by the derivative method, is applied to search and to correctly locate the peaks. Usually a number between 20 and 25 d values is required;

- among the observed $\{d_{\mathbf{h}}\}$, a selection of some base lines is performed generally involving low 2θ region of the powder diffraction pattern;
- trial values of the (hkl) Miller indices are assigned to the base lines;
- the analysis starts in the cubic symmetry, and in a stepwise manner, tests for lower symmetries are performed;
- Eq. (4.25) is solved giving feasible cell parameters;
- since more plausible cells are usually obtained, their quality is mainly assessed by using the M_{20} figure of merit [10], described in the following.

The figures of merit are introduced in the indexing software to evaluate the physical plausibility of the feasible cell and its agreement with the observed pattern, as well as to select the best cell among the possible ones. The most commonly used figure of merit is the M_{20} where only the first twenty peaks (the lowest 2θ angle peaks) of the diffraction pattern are involved in the calculations. This number is a compromise (coming from experience) between dealing with a sufficient number of observed peaks (depending on the number of unit cell parameters to be determined) and avoiding the involvement of high angle peak positions which are more affected by errors. Different figures of merit, developed to overcome the limits of M_{20}, have been proposed and implemented in some indexation programs (see [11–16]).

As final considerations, the use of $\{d_{\mathbf{h}}\}$ values in Eq. (4.25) makes precision and accuracy in detecting peak positions an essential condition for a successful indexation process. Unfortunately, in the case of powders, this condition is not always achievable due to peak overlap, bad peak resolution, 2θ zero shift, errors in measurement, low peak/background ratio.

Another source of failure in obtaining the correct result is the possible presence of impurity lines in the experimental pattern, coming from more than one chemical phase in the compound under examination.

4.2.2 Space group determination

Once the unit cell parameters have been determined, the next step in the *ab initio* structure solution pathway is the space group (Section 4.1.2) determination. The analysis of the diffraction intensities provides information on the systematically absent reflections, from which the list of space groups compatible with the determined unit cell can be obtained. This is a traditional approach, almost automatic in the case of a single crystal, often difficult and sometime ambiguous for polycrystalline compounds. For the latter, because of *powder problems*, the experimental diffraction pattern cannot be unambiguously interpreted; *e.g.*, the intensity of reflections corresponding to small peaks cannot be accurately estimated.

In the last few years, new innovative procedures, avoiding the manual inspection of the powder pattern and based on a statistical analysis of the reflection integrated intensities, have been developed for determining the space group automatically [17–20].

As an example of a space group determination method, the procedure implemented in the *EXPO* program is described. It consists of the following steps:

- a full pattern decomposition, for extracting the integrated intensities, is performed in the largest Laue symmetry compatible with the crystal system (already identified) and with no extinction conditions (*i.e.*, space group $P2/m$, $Pmmm$, $P4/mmm$, $P6/mmm$ or $Pm\text{-}3m$, in cases of monoclinic, orthorhombic, tetragonal, hexagonal or cubic symmetry, respectively);
- the intensities are submitted to a statistical analysis for the determination of the space group symmetry;
- for each extinction group compatible with the defined crystal system, a probability value is calculated as the product of the probabilities of each symmetry element regarding the group. For example, if a 2_1 symmetry element along the $\boldsymbol{a}$ axis is present, its probability is: $p(2_{1[100]}) = 1 - \langle z_{h00}\rangle_{h=2n+1}$, where $\langle z_{h00}\rangle$ is the average of the intensities of reflections $(h00)$ with h odd. The meaning of this probability is: the smaller the $\langle z\rangle$ value of a class of reflections, the larger the probability of the symmetry element responsible for the extinction of that class;
- all the feasible extinction groups are graphically ranked according to a suitable figure of merit expected to be maximum for the correct extinction group;
- the space groups (one or more) corresponding to the most probable extinction group are considered and processed in the structure solution step.

Due to unavoidable errors on the integrated intensity estimates, not always the largest figure of merit extinction symbol suggested by *EXPO* is the correct one. In order to overcome this limit, some graphical tools have been introduced in the program to help the user to easily recognize and discard wrong extinction symbols, so making successful the application of the program.

4.2.3 Profile decomposition and intensity extraction

After the unit cell parameters and the space group have been determined, the next crucial step is the profile decomposition process devoted to extract the individual integrated intensities from the experimental diffraction pattern. In spite of advances in terms of experimental devices (*e.g.*, synchrotron radiation) and computing methods, the unavoidable *powder problems* make this fundamental step not a trivial task.

The accuracy of the extracted integrated intensity values depends on the quality of the data as well as on the mathematical algorithm used for the experimental profile decomposition and strongly compromises the success probability of solving crystal structures by using the so called 'traditional' approaches like direct methods.

The process assumes that the observed experimental profile can be mathematically described by means of a calculated profile. The calculated profile intensity at the i-th step is given by:

$$y_i(\text{calc}) = s\sum_{\mathbf{h}} y_{i,\mathbf{h}}(\text{calc}) + y_i(\text{b}) = s\sum_{\mathbf{h}} m_{\mathbf{h}}\text{LP}_{\mathbf{h}} t\,|F_{\mathbf{h}}|^2 P_{i,\mathbf{h}} A_{i,\mathbf{h}} + y_i(\text{b}), \tag{4.26}$$

where s is a scale factor; the summation is over the reflections contributing to the profile intensity at the i-th step; $P_{i,\mathbf{h}}$ and $A_{i,\mathbf{h}}$ are the peak shape and asymmetry function, respectively, corresponding to the peak associated to the **h** reflection and evaluated at the i-th step; $y_i(\text{b})$ is the background contribution described by a polynomial function. The quantity $m_{\mathbf{h}}\text{LP}_{\mathbf{h}}|F_{\mathbf{h}}|^2$ represents the estimated integrated intensity of the **h** reflection ($m_{\mathbf{h}}$ is the reflection multiplicity and $\text{LP}_{\mathbf{h}}$ is the Lorentz-polarization factor).

At the present time there are two strategies widely used to determine the Bragg peak intensities from powder diffraction data: the Pawley [21] and the Le Bail [22] method, both derived from the Rietveld method [23]. The Pawley approach is a nonlinear least-squares procedure that, besides the profile and lattice parameters, refines all the individual integrated intensities. In the case of powders, correlation among the intensity values of reflections in strong overlap can provide ill-conditioning least-squares so leading to the appearance of negative intensities. The problem has been limited by modifying the method by introducing a Bayesian probability approach, able to provide positive intensities that can be successfully used for structure solution.

A different approach is given by the Le Bail method, which is an iterative application of the original Rietveld decomposition formula to observed structure factor magnitude evaluation.

In the algorithm, a good approximation of the observed integrated intensities is obtained by using the following equation:

$$I_{\mathbf{h}}(\text{obs}) = \sum_i \{y_i(\text{obs}) - y_i(\text{b})\} * y_{i,\mathbf{h}}(\text{calc})\Big/\sum_{\mathbf{k}} y_{i,\mathbf{k}}(\text{calc}),$$

where the summation at the numerator is over the peak range, $y_i(\text{obs})$ is the experimental count at $2\theta_i$ position, $y_{i,\mathbf{h}}(\text{calc})$ and $y_i(\text{b})$ as given in (4.26).

Since no structural model is available at this step, the peak intensities in $y_{i,\mathbf{h}}(\text{calc})$ are initially set to an arbitrary value and the algorithm is iteratively applied. The method is highly convergent, provides positive intensity values but tends to equiportion the peak intensity of a group of strongly overlapping reflections.

Both the Pawley and Le Bail methods are combined with a profile fitting technique based on least-squares which minimizes the difference between the observed and the calculated profiles.

To evaluate the efficiency of the powder decomposition process the R_p profile reliability parameter can be used, even if its low value is a necessary but not sufficient condition for a reliable extraction:

$$R_p = \sum_i |y_i(\text{obs}) - y_i(\text{calc})| \Big/ \sum_i y_i(\text{obs}).$$

The summation is extended over the number of experimental profile counts.

4.2.4 Structure solution

The crystal structure solution process by powder diffraction data can be carried out by using different approaches:

- The *ab initio* reciprocal space (RS) methods.
  After the steps for the unit cell parameters and the space group determination have been performed, the RS methods can be executed to activate the structure solution pathway. They require, in addition to the identified cell parameters and space group, the information concerning the experimental pattern (already used in the previous steps) and the chemical formula. The RS methods work in the reciprocal space (that of the reflections) by following a two-step approach. In the first step, the integrated intensities are extracted from the experimental powder diffraction pattern by using the Le Bail or Pawley method, in order to provide the structure factor modulus $|F_{\mathbf{h}}|$ associated to each **h** reflection; in the second step, the phase $\varphi_{\mathbf{h}}$ of structure factor $F_{\mathbf{h}} = |F_{\mathbf{h}}| \exp(i\varphi_{\mathbf{h}})$ is estimated. In crystallography, the *phase problem* has the following meaning: after an X-ray diffraction experiment, only the structure factor moduli are obtained while the relative phases are lost. They can be recovered by using the phasing methods among which the most known are: direct [2], Patterson [24–26] and maximum entropy methods [27]. Some considerations on the applications of direct methods to powder data are reported below.
- The direct space (DS) methods.
  The DS methods have become very popular in recent years and have been successfully applied for solving crystal structures, provided the expected molecular geometry is known. Avoiding the extraction of the integrated intensities from the experimental powder pattern, a so critical step for the success of the RS methods, they can succeed in cases of low resolution and/or complex structure and/or bad quality data, for which, on the contrary, RS methods often fail. The non-*ab initio* approaches work in direct space where trial structure models, built by using prior information about the expected molecular geometry, are adjusted in terms of position, orientation and internal conformation in order to find the best set of structural parameters which minimize a cost function (*CF*), usually the agreement between the observed and calculated diffraction profile. The structure solution is then equivalent to explore an N-dimensional hypersurface (where N is the number of structural parameters to be varied) to search for the minimum by using

global optimization algorithms like: grid search [28], Monte Carlo [29–31], simulated annealing (SA) [32], genetic algorithm [33–35]. It is generally assumed that the correct solution corresponds to the trial structure associated with the lowest *CF*. A description of the simulated annealing algorithm, implemented in the *EXPO* program, is given in the following.

- The hybrid methods.
  Developed and implemented in some computing programs more recently than the two previously mentioned solution approaches, the hybrid methods [36–38] combine the global optimization techniques with the phasing procedures, trying to exploit the best features of the two methods. Of particular interest is the combination of non-*ab initio* techniques with direct methods results, giving rise to a couple of hybrid procedures implemented in the *EXPO* program: *POLPO1* [39] and *POLPO2* [40]. The main features of *POLPO1* will be described below.

Direct methods solution

The solution process of powder structure by DM is carried out following the single crystal scheme (Section 4.1.3), but with some limits. In fact the difference with respect to the single crystal case is immediate: in that case the three-dimensional diffraction pattern consists of several spots, one for each reflection and the $I_{\mathbf{h}}$ quantity ($\propto |F_{\mathbf{h}}|^2$) corresponding to each reflection is easily derived. In the case of powder, the one-dimensional diffraction profile, compromised by unavoidable *powder problems*, means that the extracted structure factor moduli values (Section 4.2.3) are not accurately evaluated. Since DM are very sensitive to the reliability of the observed moduli, some problems may occur:

- the efficiency of the statistic and probabilistic calculations of DM is reduced so depressing the full phasing process. In other words, the probabilistic evaluation of the phases may be not fully reliable (the result of phasing is often affected by not negligible errors) and/or the identification of the best phasing set, by means of the largest CFOM figure of merit, may be not successful;
- DM generally end with a partial structural model, corresponding only to a rough approximation of the final structure: some atoms are imperfectly located, others are in wrong positions and/or some atoms are missing. This makes the optimization of the structural model mandatory before running the final Rietveld refinement, owing to the modest convergence rate of this technique (Section 4.2.5).

It is noteworthy that the inefficiency of DM strongly depends on the experimental resolution of the diffraction profile and/or on the structure complexity. This means that the solution process is quite trivial when an atomic resolution experimental profile, consisting of well-resolved peaks (*i.e.*, by synchrotron radiation) and corresponding to a structure of not more than 30 atoms in the asymmetric unit, is available. Recently, several methods have been developed for overcoming the limits mentioned above. They have been promptly transformed into computing algorithms and implemented in software packages increasing the probability of success of the solution process.

For example, in the *EXPO* program, procedures have been introduced devoted to

- activating, in a completely automatic way, the exploration of all the sets of phases generated by DM;
- completing and optimizing the structure model (for a general overview see [41–46]). Two of them, the WLSQ-FR and the RBM procedures, are described in the following two paragraphs, respectively.

Weighted least-squares Fourier recycling (WLSQ-FR)

As for single crystal data, in the case of powders, least-squares procedure can be used for improving an approximate structure model in order to make it more suitable for the final Rietveld refinement. The quantity to be minimized is the following:

$$\sum_{\mathbf{h}} w_{\mathbf{h}} \left[I_{\mathbf{h}}(\text{obs}) - I_{\mathbf{h}}(\text{calc}) \right]^2, \tag{4.27}$$

where $I_{\mathbf{h}}(\text{obs})$ and $I_{\mathbf{h}}(\text{calc})$ are the integrated intensities extracted from the experimental pattern for the **h** reflection and calculated by the current model, respectively. The summation is over the total number of reflections in the experimental pattern and $w_{\mathbf{h}}$ represents a weight associated to each reflection to provide a reliability estimate of $I_{\mathbf{h}}(\text{obs})$.

In the case of powder, because of the well-known problems, the extracted individual integrated intensities are not very well determined, so making the use of Eq. (4.27) not always fully efficient. The problem has been faced by using, in the *EXPO* program, a suitable weighting scheme for $w_{\mathbf{h}}$ in Eq. (4.27) [41].

The weight $w_{\mathbf{h}}$ is able to take into account the errors on the integrated intensity estimates of the overlapping reflections and to drive the refinement towards effective convergence. No matter what the specific expression of $w_{\mathbf{h}}$ is, it depends on two factors, $w_{1\mathbf{h}}$ and $w_{2\mathbf{h}}$ where $w_{1\mathbf{h}}$ hinders that least-squares are dominated by the largest intensities while $w_{2\mathbf{h}}$ takes into account the major difficulty to carefully model peak overlapping and background at high angle regions of the experimental pattern. During the procedure, to increase the possibility of converging towards the correct model, observed intensities corresponding to clusters of overlapped reflections are partitioned proportionally to the calculated intensity of each reflection in the cluster. The effect of the partition strongly depends on the quality of the model to be improved. Cycles of least-squares minimization followed by $(2|F_o| - |F_c|)$ ($|F_o|$ and $|F_c|$ are the observed and calculated structure factor moduli, respectively) Fourier map calculations are able to discard false atomic positions and move misplaced positions towards the true ones.

Resolution Bias Correction Algorithm (RBM)

It is well known that the limited experimental data resolution (RES, the smallest interplanar distance value) modifies the Fourier synthesis, introducing severe series truncation errors. In general, the worse the experimental resolution the less accurate the Fourier map: the map is negative in some regions, peaks are broadened and surrounded by positive and negative ripples of decreasing amplitude, peaks are misplaced from their correct positions and have distorted intensities. In the case of powder, diffraction atomic resolution is seldom attained because peak overlapping limits the efficiency of the decomposition process. In addition, for organic compounds, due to their rapid scattering factor decay, RES is usually far away from the atomic one (RES > 1 Å). Three different kinds of strategies, able to correct the electron density map by typical resolution bias errors, and based on Fourier map calculations, have been implemented in *EXPO* to complete and optimize the structure models. The first works in the direct space and directly modifies the electron density map [43, 44], the second operates in the reciprocal space by modifying the classical atomic scattering factors [45] and the third acts in both direct and reciprocal space [46]. The RBM procedure working in direct space has been chosen as default in the *EXPO* program. It consists of the following steps.

- Five blocks are carried out.
- In each block a $(2|F_o| - |F_c|)$ Fourier map is computed.
- One atom is positioned at the origin of the unit cell and the corresponding structure factors are calculated and used as coefficients for a Fourier map calculation. The electron density map so obtained is constituted by the main peak, located at the origin, and by ripples. The ratios $I_R(i)/I_P(0)$ are calculated where $I_P(0)$ is the main peak maximum value and $I_R(i)$ is the intensity of the i-th pixel belonging to the ripple domain.
- The quantities calculated at the previous point are used for correcting the DM map as follows: for each j-th peak in the map, the ratios $I_R(i)/I_P(0)$ are multiplied by the maximum value of the j-th peak and subtracted from the map.
- The peak is described by a best fitting Gaussian function. The map so modified is more reliable, with peaks moved on more correct positions and intensities better resolved.
- A peak search in the corrected map is executed for providing the structure model, whose fractional coordinates and the isotropic thermal factors are then optimized by least-squares.
- The final model, chemically interpreted, is graphically provided.

To obtain the complete solution, usually only one application of the default RBM run is enough; if the final model is not complete, the procedure can be applied more times.

The DS methods: Simulated annealing procedure

A simulated annealing (SA) procedure has been implemented in *EXPO*, in order to find the crystal structure solution when the *ab initio* approach does not succeed. Many useful tools are graphically available.

Two cost functions (*CF*) are widely considered:

– the agreement factor usually used in the Rietveld refinement:

$$R_{wp} = \sqrt{\frac{\sum_i w_i \left(y_i(\text{obs}) - y_i(\text{calc})\right)^2}{\sum_i w_i y_i^2(\text{obs})}},$$

where $y_i(\text{obs})$ and $y_i(\text{calc})$ have been defined in Section 4.2.3 and $w_i = 1/y_i(\text{obs})$.

– R_I factor, comparing the experimental integrated intensities $I_{\mathbf{h}}(\text{obs})$ and the intensities $I_{\mathbf{h}}(\text{calc})$ calculated by the current model:

$$R_I = \frac{\sum_{\mathbf{h}} \left|I_{\mathbf{h}}(\text{obs}) - I_{\mathbf{h}}(\text{calc})\right|}{\sum_{\mathbf{h}} I_{\mathbf{h}}(\text{obs})}.$$

Before starting SA, the experimental integrated intensities are preliminary and automatically extracted by *EXPO* by using the Le Bail algorithm. The advantage of adopting R_I (instead of R_{wp}) is the significant reduction of computational time; on the contrary, R_{wp} is preferred if the overlap is severe and consequently the $I_{\mathbf{h}}(\text{obs})$ values can be affected by large errors.

The starting model can be managed by *EXPO* in different formats: MDL Molfile (*.mol), MOPAC file (*.mop), Tripos Sybyl file (*.mol2,*.ml2), CIF file (*.cif), Protein Data Bank file (*.pdb), XYZ format (*.xyz), Fenske–Hall Z-Matrix (*.fch). Different types of polyhedra (tetrahedron, octahedron, square plane) can be imported by *EXPO*. During each SA run, observed, calculated and difference profile, together with the current structure model, are automatically visualized each time a new configuration, with a lower *CF* value, is found. A plot of the *CF* value depending on the SA moves is also provided.

During each SA run, *EXPO* changes the torsion angles (automatically identified by the program), the orientation and the position of the model. Since distances and angles of the trial model are not varied, wrong results can be obtained if they are consistently different from the target model. Another tool is represented by the dynamical occupancy correction [47], very useful when DS methods are applied to non-molecular crystals for which some atoms are expected to occupy special positions and different building blocks can share one or more atoms.

At the end of the automatic procedure the best 10 solutions corresponding to the lowest *CF* values, are saved in CIF files and graphically shown to the user.

The hybrid approaches: The *POLPO1* procedure

When applied to heavy atom structures, DM often provide an electron density map far from the true one: the heavy atoms are usually well located while the light atoms are missing or severely misplaced. The limit can be overcome, in the case of inorganic structures with known heavy atom polyhedral coordination (tetrahedral and/or octahedral), *via* the application of *POLPO1* procedure, which is able to complete the partial model starting from few correctly located heavy atoms. It is a combination of DM results (the positions of the cations) with a Monte Carlo approach (to locate the anions), exploiting the prior information about the polyhedral coordination of the heavy atoms and, for each polyhedron, the expected cation-anion bond distances and angles. In a cyclic procedure, couples of cations with a distance compatible with a bridge anion are selected and, around them, the polyhedra are geometrically built and freely rotated in positions not violating the crystal chemical rules. Some checks are executed on the calculated feasible anions to evaluate their reliability. For each cation, the expected coordination of its polyhedron is verified. Once the polyhedra of the selected cations have been completed, the process is iterated by choosing a new pair of cations. The procedure stops when all the anions have been located.

The reliability of each feasible structure model is assessed by using the agreement parameter comparing observed and calculated profiles (R_{wp}). At the end of the *POLPO1* run, the ten models corresponding to the smallest R_{wp} values are selected and checked in terms of chemical consistency.

Two kinds of problems can mainly hinder the success of the procedure:

- severely misplaced heavy atom DM positions: they may lead to false cation connectivity;
- strong polyhedral distortions in the target model. Even if the procedure generates regular polyhedra, the use of tolerances on distances and angles can also create distorted polyhedra.

A generalization of the described procedure is represented by the *POLPO2* approach, also implemented in the *EXPO* program and able to complete inorganic structures when only some of the heavy atoms are positioned, *i.e.*, some cations and all the light atoms are missing.

4.2.5 Rietveld refinement

In the structure solution process by powder diffraction data, the last step is the model refinement. To refine crystal structures from powder diffraction data, the largely used least-squares based approach is the Rietveld method, minimizing the following quantity:

$$\sum_i w_i \left[y_i(\text{obs}) - y_i(\text{calc})\right]^2,$$

where the summation is over the number of counts in the experimental diffraction pattern.

In this approach, the profile and the structural model parameters, *i.e.*, atomic coordinates, thermal displacements, site occupancies, unit cell, peak shape, *etc.*, are refined. In the Rietveld method the relationships between the adjustable parameters and the calculated counts are non-linear: a false minimum can be reached by least-squares. As a consequence, a correct solution can be obtained if the starting model is sufficiently close to the target structure. For powder diffraction data, because of the simultaneous effects of *powder problems*, the convergence by Rietveld technique is not always attained.

In the *EXPO* program, a user-defined Rietveld refinement strategy, supported by a user-friendly graphic interface, has been introduced. The quality of the refinement process and its progress can be assessed by using some figures of merit like [48]:
- R_p and R_{wp}, the unweighted and weighted profile R-factors calculated by the full pattern;
- R_p', R_{wp}' the corresponding background-subtracted factors;
- R_{exp} the statistically expected R value;
- $\chi^2 = (R_{wp}/R_{exp})^2$;
- the Durbin–Watson d-statistic.

In addition, during the refinement, restraints on bond distances, angles and planes can be applied to the models if the molecular geometry is known *a priori*. Graphical tools are available for visualizing
- the observed, calculated and difference patterns and the cumulative χ^2 value;
- the evolution of the structure model during the refinement in real time.

4.2.6 Examples

Two examples of structure solution by powder diffraction data, by using the *EXPO* program, will be described. The first is related to a default *ab initio* solution process, the second to the model optimization of an organic structure.

Example 1. This concerns the cerium (IV) oxide Cr_4O_8 simple structure [49]: cubic, *Fm-3m*, 2θ experimental range (20–110)°, 12 reflections, *RES* = 0.94 Å; medium quality laboratory data, whose experimental pattern is shown in Figure 4.13, 1 Ce and 1 O atoms in the asymmetric unit. It is a crystal nanostructure with average crystallite dimension of about 15 nm.

To run the program, an input file must be prepared, supplying the minimum information constituted by
- the ASCII filename containing the observed diffraction pattern;

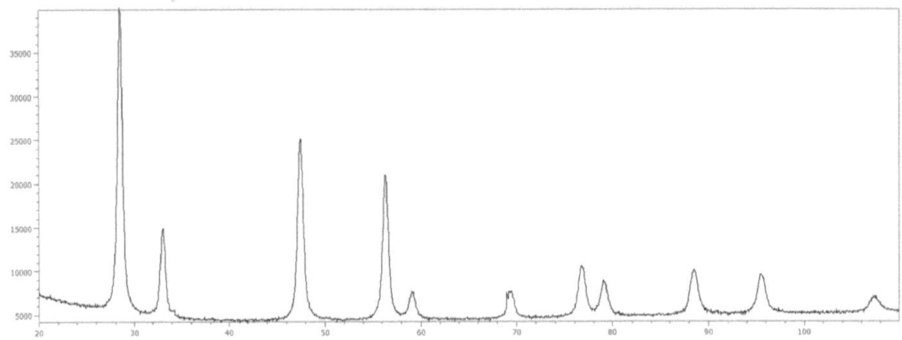

Fig. 4.13: Cerium (IV) oxide, Cr_4O_8. The measured diffraction profile.

- the wavelength (in Å) used to collect the data (X-rays, neutron and synchrotron data can be treated);
- the chemical unit cell content;
- the instructions to execute, in sequence and automatically, all the solution steps.

This last point concerns the following.

- *Unit cell parameter determination.* It is carried out by using the indexation program *N-TREOR09* [16], implemented in the *EXPO* software and following the procedures described in Section 4.2.1. A preliminary automatic peak search procedure is able to detect peaks in the pattern. Vertical bars above the peaks show and mark them: 11 observed peaks, over all the experimental range, are selected and used to find feasible cells. At the end of the indexing procedure, *N-TREOR09* suggests two plausible cells (Fig. 4.14a), graphically shown to the user. The first one, that with the largest M_{20} figure of merit, is the cubic cell with parameters $a = b = c = 5.41266$ Å, $\alpha = \beta = \gamma = 90°$. Its M_{20} value equal to 196 is much larger that the value of the figure of merit (= 18) associated to the other hexagonal plausible cell. We select the cubic cell.
- *Space group determination.* After the cell suggested by *N-TREOR09* as the correct one has been graphically selected by the user, the space group determination process automatically starts following the procedure described in Section 4.2.2. In the case of cerium (IV) oxide, at the end of the procedure, 18 possible extinction symbols corresponding to the cubic crystal system, listed according to the probability values, are supplied: (*F*---) is the most probable extinction group. Five space groups (*F*23, *Fm*-3, *F*432, *F*-43*m* and *Fm*-3*m*) correspond to it (Fig. 4.14b); the solution process should be attempted by using all the possible space groups. We prefer to use the space group *Fm*-3*m* for the first attempt as a suggestion coming from information regarding structures already solved, reported in literature and similar to our structure under study.

- *Decomposition of the experimental diffraction profile and intensity extraction.* The step, executed according to the Le Bail method as described in Section 4.2.3, ends with a profile reliability parameter quite small ($R_p = 15.5\,\%$) assessing a good agreement between the observed and the calculated profile (Fig. 4.14c). The single intensities corresponding to each one of the 12 reflections in the full pattern are evaluated. From them the structure factor moduli values are derived.
  Two reflections corresponding to the 2θ angular position of 95.39285° are systematically overlapped (Fig. 4.14c).
- *Structure solution.* DM are automatically applied following the steps described in Section 4.1.3. The set of phases corresponding to the largest CFOM is automatically selected and submitted to the WLSQ-FR procedure described in Section 4.2.4 giving, at the end, the correct model: 2 atoms are rightly labeled and located in the asymmetric unit (*a posteriori* analysis reveals that the two atom positions are close to the true ones with a mean distance $\langle d \rangle = 0.001\,\text{Å}$). The final model in the unit cell is shown in Figure 4.14d.

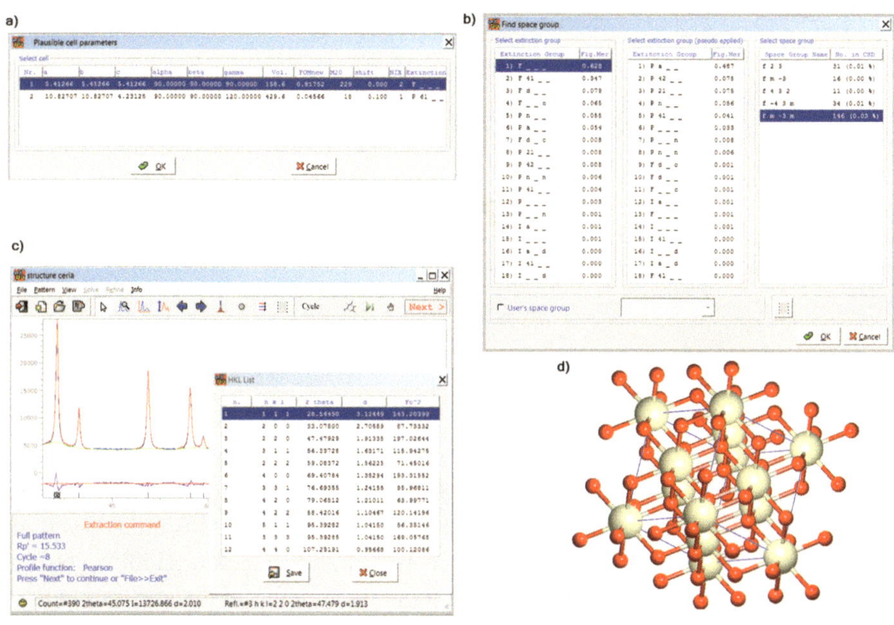

Fig. 4.14: Cerium (IV) oxide, Cr_4O_8. The results by *EXPO* of each step of the structure solution process are shown: (a) indexation; (b) space group determination; (c) profile decomposition and intensity extraction; (d) structure model (in the unit cell).

Example 2. In the case of organic compounds, *EXPO* automatically applies the RBM procedure described in Section 4.2.4 for correcting the electron density map by errors due to limited *RES* and providing the solution. The example concerns the phenylacetic acid (pharmaceutical compound) $C_8H_8O_2$ structure [50]: monoclinic, $P2_1/a$, 2θ experimental range (5–65)°, 258 reflections, *RES* = 1.43 Å; medium quality laboratory data; two O and eight C atoms in the asymmetric unit. The first phasing set (ranked according to CFOM), automatically selected by the program for the Fourier map calculation, does not supply the correct solution which is instead provided by the fifth one (as obtained by the automatic procedure exploring all the phasing trials). At the end of the DM process, the obtained Fourier map is given (Fig. 4.15a): it is completely uninterpretable (*a posteriori* analysis reveals that seven positions are close to the true ones with distances less than 0.6 Å). The final model by RBM is shown in Figure 4.15b: it is strongly improved and provides the solution.

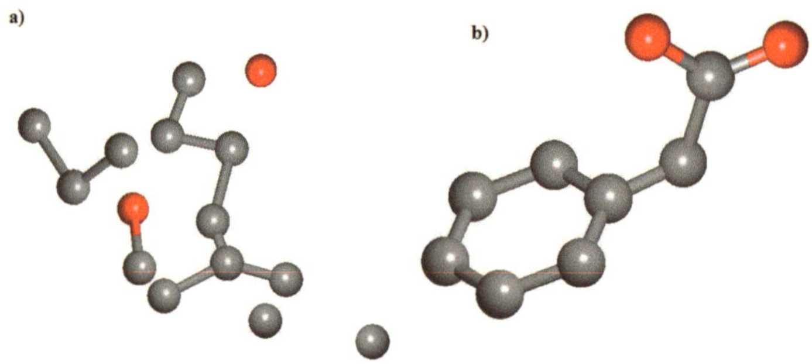

Fig. 4.15: Phenylacetic acid $C_8H_8O_2$: (a) DM structure model; (b) model optimized by the automatic *EXPO* RBM procedure.

4.3 Conclusions

X-ray diffraction is a powerful tool for determining the crystal structure of small molecules. Software with sophisticated graphical and computing performance is available for automatically solving the structure. The solution process is quite trivial for single crystal, much more complex for microcrystalline and nanocrystalline powder. The great progress achieved in the last twenty years, in terms of theories, computing strategies, experimental devices and, particularly, software packages has strongly improved the solution process by powder, so reducing the gap with respect to the single crystal case. *EXPO* is a widely used computer program able to automatically carry out all the powder solution steps, from indexing to structure model refinement. It represents a very useful tool for solving organic, inorganic and metallorganic structures, provided that the structure complexity is not very large (about 40 non-hydrogen atom in the asymmetric unit) and/or the experimental data resolution not very bad (better that 1.5 Å).

Bibliography

[1] Giacovazzo C. The diffraction of X-rays by crystals, in: Giacovazzo C, Monaco HL, Artioli G et al. (eds.), Fundamentals of Crystallography, 3rd ed. pp. 157-234, Oxford, England: IUCr/Oxford University Press, 2011.
[2] Giacovazzo C. Direct Phasing in Crystallography. Oxford, England: IUCr/Oxford University Press, 1998.
[3] Wilson AJC. Determination of absolute from relative intensity data. Nature, 1942; 150: 151–2.
[4] Cochran W. Relations between the phases of structure factors. Acta Cryst 1955; 8: 473–8.
[5] Karle J, Hauptman HA. A theory of phase determination for the four types of non centrosymmetric space groups 1P222, 2P22, 3P12, 3P22. Acta Cryst. 1956; 9: 635–51.
[6] Boufas S, Hadjadj N-E, Merazig H, Moliterni AG, Altomare A. Hydrogen bonding and structure of $Ba_2Ru_2Cl_{10}O.10H_2O$. Acta Cryst. 2007; C63: i60–2.
[7] Dinnebier RE, Billinge SJL (eds.). Powder Diffraction Theory and Practice. Cambridge, England: RSCPublishing, 2008.
[8] Altomare A, Camalli M, Cuocci C, Giacovazzo C, Moliterni A, Rizzi R. EXPO2009: Structure solution by powder data in direct and reciprocal space. J. Appl. Cryst. 2009; 42: 1197–202.
[9] Methods, Problems and Solutions. Diffraction Indexing including a Literature Search List. (Accessed July 6, 2012, at www.ccp14.ac.uk/solution/indexing).
[10] de Wolff PM. A simplified criterion for the reliability of a powder pattern indexing. J. Appl. Cryst. 1968; 1: 108–13.
[11] Smith GS, Snyder RLJ. F_N: A criterion for rating powder diffraction patterns and evaluating the reliability of powder pattern index. Appl. Cryst. 1979; 12: 60–5.
[12] Ishida T, Watanabe YJ. Probability computer method of determining the lattice parameters from powder diffraction data. Phys. Soc. Japan. 1967; 23: 556–65.
[13] Ishida T, Watanabe YJ. Analysis of powder diffraction pattern of monoclinic and triclinic crystals. Appl. Cryst. 1971; 4: 311–6.
[14] Bergmann JZ. EFLECH/INDEX- Another try of whole pattern indexing. Kristallogr. Suppl. 2007; 26: 197–202.
[15] Le Bail A. Structure Solution, in: Clearfield A, Reibenspies JH, Bhuvanesh N (eds.), Principles and Applications of Powder Diffraction, pp. 261–309, Oxford, England: Wiley-Blackwell, 2008.
[16] Altomare A, Campi G, Cuocci C et al. Advances in powder diffraction pattern indexing: *N-TREOR09*. J. Appl. Cryst. 2009; 42: 768–75.
[17] Markvardsen AJ, David WIF, Johnson JC, Shankland K. A probabilistic approach to space group determination from powder diffraction data. Acta Cryst. 2001; A57: 47–54.
[18] Altomare A, Caliandro R, Camalli M et al. Space-group determination from powder diffraction data: A probabilistic approach. J. Appl. Cryst. 2004; 37: 957–66.
[19] Altomare A, Camalli M, Cuocci C et al. Space group determination: improvements in *EXPO2004*. J. Appl. Cryst. 2005; 38: 760–7.
[20] Altomare A, Camalli M, Cuocci C, Giacovazzo C, Moliterni AGG, Rizzi R. Advances in space-group determination from powder diffraction data. J. Appl. Cryst. 2007; 40: 743–8.
[21] Pawley GS. Unit-cell refinement from powder diffraction scans. J. Appl. Cryst. 1981; 14: 357–61.
[22] Le Bail A, Duroy H, Fourquet JL. Ab-initio structure determination of LiSbWO by X-ray powder diffraction. Mater. Res. Bull. 1988; 23: 447–52.
[23] Rietveld HM. A profile refinement method for nuclear and magnetic structures. J. Appl. Cryst. 1969; 2: 65–71.

[24] Rius J, Crespi A, Torrelles X. A direct phasing method based on the origin-free modulus sum function and the FFT algorithm. XII. Acta Cryst. 2007; A63: 131–4.
[25] Rius J, Frontera CJ. Application of the constrained S-FFT direct-phasing method to powder diffraction data. XIII. J. Appl. Cryst. 2007; 40: 1035–8.
[26] Rius J. Patterson-function direct methods for structure determination of organic compounds from powder diffraction data. XV. Acta Cryst. 2011; A67: 63–7.
[27] Gilmore C, Dong W, Bricogne G. A multisolution method of phase determination by combined maximization of entropy and likelihood. VI. The use of error-correcting codes as a source of phase permutation and their application to the phase problem in powder, electron and macromolecular crystallography. Acta Cryst. 1999; A55: 70–83.
[28] Chernyshev VV, Schenk H. A grid search procedure of positioning a known molecule in an unknown crystal structure with the use of powder diffraction data. Z. Krist. 1998; 213: 1–3.
[29] Harris KDM, Tremayne M, Lightfoot P, Bruce PG. Crystal structure determination from powder diffraction data by Monte Carlo methods. J. Am. Chem. Soc. 1994; 116: 3543–7.
[30] Andreev YG, Lightfoot P, Bruce PG. A general Monte Carlo approach to structure solution from powder diffraction data: Application to poly(ethylene oxide)$_3$:LiN(SO_3CF_3)$_2$. J. Appl. Cryst. 1997; 30: 294–305.
[31] Tremayne M, Kariuki BM, Harris KDM, Shankland K, Knight KS. Crystal structure solution from neutron powder diffraction data by a new Monte Carlo approach incorporating restrained relaxation of the molecular geometry. J. Appl. Cryst. 1997; 30: 968–74.
[32] Kirkpatrick S. Optimization by simulated annealing: quantitative studies. J. Stat. Phys. 1983; 34: 975–86.
[33] Goldberg DE. Genetic Algorithms in Search, Optimization, and Machine Learning. New York, USA: Addison-Wesley, 1989.
[34] Kariuki BM, Serrano-González H, Johnston RL, Harris KDM. The applications of a genetic algorithm for solving crystal structures from powder diffraction data. Chem. Phys. Lett. 1997; 280: 189–95.
[35] Shankland K, David WIF, Csoka T. Crystal structure determination from powder diffraction data by the application of a genetic algorithm. Z. Krist. 1997; 212: 550–2.
[36] Johnston JC, David WIF, Markvardsen AJ, Shankland K. A hybrid Monte Carlo method for crystal structure determination from powder diffraction data. Acta Cryst. 2002; A58: 441–7.
[37] Brenner S, McCusker LB, Baerlocher C. Using a structure envelope to facilitate structure solution from powder diffraction data. J. Appl. Cryst. 1997; 30: 1167–72.
[38] Altomare A, Caliandro R, Giacovazzo C et al. Direct methods and simulated annealing: a hybrid approach for powder diffraction data. J. Appl. Cryst. 2008; 41: 56–61.
[39] Altomare A, Giacovazzo C, Guagliardi A, Moliterni AGG, Rizzi, R. Completion of crystal structures from powder data: The use of the coordination polyhedra. J. Appl. Cryst. 2000; 33: 1305–10.
[40] Giacovazzo C, Altomare A, Cuocci C, Moliterni AGG, Rizzi R. Completion of crystal structures from powder data: A new method for locating atoms with polyhedral coordination. J. Appl. Cryst. 2002; 35: 422–9.
[41] Altomare A, Cuocci C, Giacovazzo C, Moliterni AGG, Rizzi R. Powder diffraction: the new automatic least-squares Fourier recycling procedure in EXPO2005. J. Appl. Cryst. 2006; 39: 558–62.
[42] Altomare A, Camalli M, Cuocci C, Giacovazzo C, Moliterni AGG, Rizzi R. Direct methods and the solution of organic structures from powder data. J. Appl. Cryst. 2007; 40: 344–8.
[43] Altomare A, Cuocci C, Giacovazzo C, Kamel GS, Moliterni A, Rizzi R. Minimally resolution biased electron-density maps. Acta Cryst. 2008; A64: 326–36.

[44] Altomare A, Cuocci C, Giacovazzo C, Moliterni A, Rizzi R. Correcting resolution bias in electron density maps of organic molecules derived by direct methods from powder data. J. Appl. Cryst. 2008; 41: 592–9.
[45] Altomare A, Cuocci C, Giacovazzo C, Moliterni A, Rizzi, R. Correcting electron-density resolution bias in reciprocal space. Acta Cryst. 2009; A65: 183–9.
[46] Altomare A, Cuocci C, Giacovazzo C, Moliterni A, Rizzi, R. The dual-space resolution bias correction algorithm: applications to powder data. J. Appl. Cryst. 2010; 43: 798–804.
[47] Cerny R & Favre-Nicolin V. FOX, 'Free objects for crystallography': A modular approach to ab initio structure determination from powder diffraction. J. Appl. Cryst. 2002; 35: 734–43.
[48] Young RA. The Rietveld Method. Oxford, England: Oxford University Press, 1995.
[49] Aresta M, Dibenedetto A, Pastore C et al. E. Cerium(IV)oxide modification by inclusion of a hetero-atom: A strategy for producing efficient and robust nano-catalysts for methanol carboxylation. Catalysis Today. 2008; 137: 125–31.
[50] Florence AJ, Shankland N, Shankland K et al. Solving molecular crystal structures from laboratory X-ray powder diffraction data with *DASH*: The state of the art and challenges. J. Appl. Cryst. 2005; 38: 249–59.

Elisa Borfecchia, Giovanni Agostini, Silvia Bordiga, Elena Groppo, Claudio Garino, Roberto Gobetto, Angelo Agostino, Lorenzo Mino, Marco Truccato, Diego Gianolio, Andrea Piovano, Gema Martinez-Criado, Luca Salassa, and Carlo Lamberti

5 Structural and electronic characterization of nanosized inorganic materials by X-ray absorption spectroscopies

5.1 Introduction

Starting from the late seventies, the progressively increased availability of synchrotron light sources has allowed the execution of experiments requiring a high X-ray flux in a continuous interval [1–4]. Among the various techniques developed, X-ray absorption spectroscopy (XAS, also known as X-ray absorption fine-structure, XAFS) [5–13] in both near (XANES) and post (EXAFS) edge regions, has become a powerful characterization technique widely employed in many research fields concerning inorganic chemistry, such as catalysis [13–17], organometallic and coordination complexes [18], metal nanoparticles [19, 20], electrochemical processes [21, 22], coordination polymers or metallorganic frameworks [23], bio-inorganic molecules including metalloproteins [24–27], chemistry of actinide elements [28], and solid state chemistry [12, 29–31].

In this chapter, we will provide a brief introduction to the basic theory of XAS spectroscopy (Section 5.2), followed by selected examples having some didactic perspective (Sections 5.3–5.7). For the sake of brevity, only a fraction of the subjects mentioned above will be discussed; in particular, the chapter will deal with: the reactivity of the $CuCl_2/Al_2O_3$-based catalysts for ethylene oxychlorination (Section 5.3); the structure, the electronic configuration and the reactivity of Cp_2Cr molecules encapsulated in porous solids (Section 5.4); the structural characterization of metal complexes in solution (Section 5.5); the structural characterization of the UiO-66 and UiO-67 class of metallorganic frameworks (Section 5.6); and the determination of the local structure of an $Al_{xw}Ga_{yw}In_{1-xw-yw}As/Al_{xb}Ga_{yb}In_{1-xb-yb}As$ strained heterostructure grown on InP by metallorganic vapor phase epitaxy (Section 5.7) as an example of space-resolved EXAFS applied to a topic pertinent to solid state chemistry.

5.2 XAS spectroscopy: Basic background

The aim of this section is to provide the reader with a concise review of the basic physical principles on which the interpretation of EXAFS and XANES data is based. For a more detailed description of the theoretical background and the experimental aspects of XAS we refer to the extensive specialized literature (*e.g.*, [5–13]).

5.2.1 Theoretical background of XAS spectroscopy

XAS measures the variations of the X-ray absorption coefficient μ as a function of the incident X-ray energy E. According to the Fermi golden rule [32, 33], the XAS signal is proportional to the electron transition probability from the core-state $|i>$ of energy E_i to the unoccupied state $|f>$ of energy E_f, as expressed by Eq. (5.1), where the product $\mathbf{e}\cdot\mathbf{r}$ indicates the electronic transition dipole operator [7], where $\rho_i(occ)$ and $\rho_f(unocc)$ are the densities of initial occupied and final unoccupied states, respectively, and where $\delta(E_f - E_i - E)$ is a Dirac delta function [12].

$$\mu(E) \propto |<i|\mathbf{e}\cdot\mathbf{r}|f>|^2 \delta(E_f - E_i - E)\rho_i(occ)\rho_f(unocc). \tag{5.1}$$

The behavior of the $\mu(E)$ function is represented in Figure 5.1b. A general decrease of the absorption with increasing incident energy can be noticed, following approximately the law [13]:

$$\mu(E)/\rho \approx Z^4/AE^3, \tag{5.2}$$

where ρ is the sample density, Z the atomic number and A the atomic mass. This equation holds for a sample containing a unique chemical species like a metal foil, but can be easily generalized for any sample of known composition. In Figure 5.1b the presence of the characteristic saw-tooth like edges is also evident, whose energy position is a distinctive features of each kind of absorbing atom. These absorption edges correspond to transitions where a core-orbital electron is excited to (i) the free continuum (*i.e.*, when the incident energy is above the ionization energy of the absorber atom) or (ii) unoccupied bond states lying just below the ionization energy. The nomenclature adopted for the edges recalls the atomic orbitals from which the electron is extracted, as shown in Figure 5.1a: K-edges are related to transitions from orbitals with the principal quantum number n = 1 ($1s_{1/2}$), L-edges refer to electron from the n = 2 orbitals (L_I to $2s_{1/2}$, L_{II} to $2p_{1/2}$, and L_{III} to $2p_{3/2}$ orbital), and so on for M, N, ... edges.

When the energy of the X-ray photon exceeds the ionization limit (case (i) mentioned above), the excited electron (generally named "photoelectron") has a kinetic energy E_K given by $E_K = h\upsilon - E_B$, where E_B indicates the electron binding energy that is typical of the absorption edge (K, L_I, L_{II} or L_{III}) of the selected atomic species [13, 34]. Once ejected, the photoelectron propagates thorough the sample as a spherical wave diffusing from the absorber atom, with a wave vector of modulus k defined by Eq. (5.3):

$$k = \frac{1}{\hbar}\sqrt{2m_e E_K}. \tag{5.3}$$

A close zoom on the energy region in proximity of an absorption edge shows a well-defined fine structure. In particular, only when the absorber is surrounded by neighboring atoms (molecules or crystals) a structure of oscillatory nature modulates the smooth $\mu(E)$ profile at energies above the edge. Figure 5.1c provides an example for the Cr K-edge spectra of Cp_2Cr in different environments and after interaction with CO (see Section 5.4.2). Such modulation in the absorption coefficient derives from the interference between the outgoing photoelectron wave diffusing from the absorber and the wave fronts backscattered by the neighboring atoms [35, 36]. In a typical XAS experiment, the energy range probed around the edge is conventionally divided into two different regions (Fig. 5.1c):

(i) X-ray absorption near edge structure (XANES) region: portions of the XAS spectrum just below and above the edge energy;
(ii) Extended X-ray absorption fine structure (EXAFS) region: portion at higher energies with respect to the edge (from tens to hundreds of eV), characterized by the oscillatory modulation in the absorption coefficient.

Hereby, we will briefly discuss the main information that can be extracted from the analysis of each of the two regions listed above.

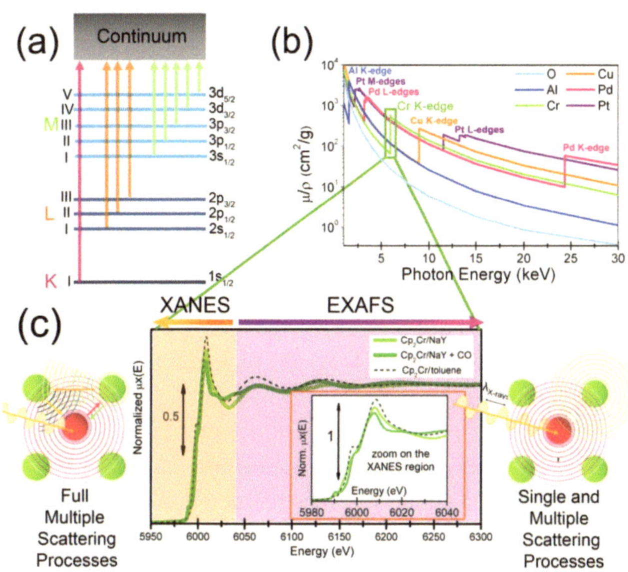

Fig. 5.1: Part (a): X-ray absorption edges nomenclature and representation of their relation with the atomic orbitals from which the electron is extracted. Part (b): General behavior of the X-ray absorption coefficient μ/ρ, see Eq. (5.2), as a function of the incident X-ray energy E for O (Z = 8), Al (Z = 13), Cr (Z = 24), Cu (Z = 29), Pd (Z = 46), and Pt (Z = 78). Data obtained from NIST website http://physics.nist.gov/PhysRefData/XrayMassCoef/tab3.html. Note the logarithmic scale of the ordinate axis. Part (c): Cr K-edge XAFS of Cp_2Cr molecule in toluene solvent (dashed line) and hosted inside the cavities of Na-Y zeolite before (light green line) and after (dark green light) interaction with CO [37]. Data collected at BM26A beamline [38] of the ESRF. The conventional division between XANES and EXAFS regions and the schematic models of full multiple and single scattering processes, dominating respectively the XANES and EXAFS region, are indicated (color code: absorber atom in magenta; neighbor atoms that backscatter the photoelectron wave outgoing from the absorber in green). Unpublished figure.

5.2.2 The XANES region

The XANES region can be further divided into three main sub-intervals: we can distinguish the pre-edge region, the absorption edge, and the post-edge region, up to 30–50 eV after the edge jump. The investigation of these three portions of the XANES spectrum and the interpretation of the different physical phenomena governing their characteristic features provide plenty of information, both of electronic and structural nature [39].

The pre-edge region can include well-defined components, related to electronic transitions from the core level to empty bound states [40], as for example the unoccupied states of the narrow d-band lying just above the Fermi energy in transition metals. The features of the pre-edge peaks are directly influenced by the density of the states (DOS) and the states occupancy, therefore their investigation is extremely helpful in unraveling the electronic structure of molecules and complexes [7, 11]. Moreover, the intensity of pre-edge peaks is governed by the selection rules for dipolar transitions, which can be relaxed depending on the local symmetry around the absorber atom [41]. Pre-edge features related to transitions that are forbidden in the purely dipolar case can however clearly appear in the XANES spectrum, hence providing information about the geometrical coordination environment of the absorber.

The energy position of the absorption edge is sensitive to the oxidation state of the absorber due to the charge-shielding effect on the ionization energy: practically, a higher oxidation number corresponds to an edge shift towards higher energies [31, 42–45]. In addition, the edge location is influenced by the coordination geometry, which ultimately is determined by the molecular orbitals' energy position [37, 46, 47].

The edge and post-edge regions are particularly sensitive to the geometrical arrangement of the neighboring atoms in the restricted local cluster around the absorber. This makes XANES a very powerful instrument for direct structural characterization in a wide range of systems, including solids [31, 45], solutions [48], surfaces [49] metal nanoparticles [50], and biological samples [51]. The relation between geometrical features and XANES response mainly depends on the multiple scattering processes (Fig. 5.1c) of the excited photoelectron after the ionization threshold, which dominate with respect to the single scattering events due to the lowered photoelectron energy. Moreover, the first 10 eV after the ionization limit are in many cases influenced also by other kinds of physical interactions, directly dependent on the properties of the investigated sample (the so-called "shape resonances" in molecules, "core excitons" in crystals, "unoccupied local electronic states" in metals and insulators [52]).

5.2.3 The EXAFS region

The EXAFS region of the spectrum is located at higher energies and is characterized by the modulation of the absorption coefficient $\mu(E)$. Such a feature is caused by the interference between the X-ray waves diffused by the absorber atom and backscattered by its neighbors. Hence, EXAFS oscillations can be related *via* Fourier transform to a specific spatial arrangement of the atoms in the local environment of the absorber, bridging the energy space (actually the k-space of the photoelectron) to the real distances R-space. This crucial point is at the basis of the EXAFS analysis procedure developed after the milestone works of Sayers, Lytle and Stern [35, 36, 53].

The higher photoelectron kinetic energy in the EXAFS region implies that the phenomenon is not dominated by the full multiple scattering regime that instead dominates in the XANES region [53]; consequently data analysis can be performed using the simpler Fourier transform operation [35].

The EXAFS signal $\chi(E)$ is generally expressed as the oscillatory part of the $\mu(E)$ function, normalized to the edge-jump, *i.e.*, $\chi(E) = [\mu(E) - \mu_0(E)]/\Delta\mu_0(E)$, where $\mu_0(E)$ is the atomic-like background absorption and $\Delta\mu_0(E)$ the normalization function that takes into account the edge jump intensity. Above the absorption edge, the energy E can be substituted with the photoelectron wave-vector k using Eq. (5.3), therefore obtaining the EXAFS function $\chi(k)$. The relation between the modulation of the $\chi(k)$ signal and the structural parameters is provided by the EXAFS formula that, in the single scattering approximation, is reported in Eq. (5.4).

$$\chi(k) = S_0^2 \sum_i N_i F_i(k) e^{-2\sigma_i^2 k^2} e^{\frac{2r_i}{\lambda(k)}} \frac{\sin[2kr_i + 2\delta_i(k) + \theta_i(k)]}{kr_i^2} \tag{5.4}$$

S_0^2 is the overall amplitude reduction factor; the index i runs over all the different shells of neighboring atoms around the absorber, $F_i(k)$ is the backscattering amplitude as a function of k for each shell, N_i is the coordination number (number of equivalent scatterers), σ_i is the Debye–Waller factor accounting for thermal and static disorder. The parameter r_i indicates the interatomic distance of the i-th shell from the central absorber. The phase shift of the photoelectron is separated in two contributions, related to the absorber ($2\delta_l$) and to the scatterer (θ_i).

In Eq. (5.4) the electron backscattering amplitude $F_i(k)$ is measured in Å [54,55]. Indeed $F_i(k)$ defines the weight that the i-th neighbor has in the overall EXAFS signal. As the photoelectron is mainly scattered by the electron clouds of the neighbor atoms, $F_i(k)$ is larger for larger Z neighbors. Consequently, as in the case of X-ray scattering, EXAFS is less efficient in the detection of low Z neighbors and the discrimination among neighbors having similar Z is critical. When the difference in Z is sufficiently large, then both backscattering amplitude $F_i(k)$ and phase shift functions are markedly different to allow an easy discrimination between the different neighbors.

The presence of the phase terms $2\delta(k) + \theta(k)$ inside the sine function is the reason why the FT of the raw $k^n\chi(k)$ functions always reports a peak in R-space at 0.2–0.4 Å shorter distance with respect to the true distance. Only a phase corrected FT results in a distances distribution in R-space centered around the correct value, *vide infra* the discussion in Section 5.3.2.1 and Figure 5.2b,c,d.

The term $\lambda(k)$ is the energy-dependent photoelectron mean free path, typically few Å [56], determining the local nature of the technique that can investigate only up to $\approx$ 5–8 Å around the photo-excited atom. This apparent limitation of EXAFS is conversely a big advantage in the investigation of disordered materials like glasses, liquids or gases as the long-range order is not required [57]. Furthermore, EXAFS provides element-selective structural information on the local environment of the absorber, whereas the response from other widespread techniques, such as X-ray diffraction, contains contributions from all the atoms included in the probed volume.

Coming to the Debye–Waller term $\exp(-2\sigma_i^2 k^2)$, it has the same physical origin as the term $\exp(-q^2 u_n^2/6)$ used to refine diffraction experiments [58, 59], where q ($q = 4\pi \sin(\theta)/\lambda$) is the scattering vector exchanged in the elastic scattering process, and u_n indicates the isotropic mean squared displacement for the n-th atom. Actually, the thermal parameters σ_i^2 measured in an EXAFS experiment and the u_n^2 parameters measured in a diffraction experiment represent two different aspects of the atomic vibrations in the probed sample. The u_n^2 parameter (or the anisotropic mean squared displacements ($u_{n,a}^2$, $u_{n,b}^2$, $u_{n,c}^2$) measures how the n-th atom vibrates inside the unit cell along **a**, **b**, and **c** directions. Similarly, the σ_i^2 parameter determined in EXAFS experiments measures how the couple of atoms constituted by the absorber and the i-th shell neighbor vibrate along the direction that connect these two atoms. This means that in cases of strongly correlated vibrations, the scatterer and the absorber can vibrate in phase, resulting in a low σ_i^2 parameter and in a larger u_n^2 parameter.

The standard EXAFS formula, Eq. (5.4), provides a convenient parameterization for fitting the local atomic structure around the absorbing atom to the experimental EXAFS data [60]. The dependence of the oscillatory structure of the EXAFS signal on interatomic distance and energy is clearly reflected in the $\sin(2kr_i)$ term. The strength of the interfering waves depends on the type and number of neighboring atoms through the backscattering amplitude $F_i(k)$ and the coordination number N_i, and hence these two parameters are primarily responsible for the magnitude of the EXAFS signal. Once the phase and amplitude functions have been independently measured on model compounds or computed using *e.g.*, *ab initio* or DFT methods, the structural parameters N_i, r_i, and σ_i^2, can be determined in a least-squares approach, where the difference between the experimental and the modeled $k^n\chi(k_j)$ function is minimized along all the sampled experimental points k_j. The minimization routine can be performed either in k-space, directly on the measured $k^n\chi(k_j)$ function, or in R-space, working on the Fourier-transformed functions. Therefore, for each coordination shell, the coordination number, the atomic distance and the thermal factor can be extracted from an accurate EXAFS study. Extending the Nyquist-Shannon theorem [61, 62] (also known as sampling theorem) to the EXAFS case, the maximum number of optimized parameters cannot exceed the number of truly independent points (n_{ind}), where n_{ind} is defined by the product of the sampled interval in k-space (Δk) and the interval in R-space (ΔR)

containing the optimized shells:

$$n_{ind} = 2\Delta k \Delta R / \pi. \tag{5.5}$$

A careful monitoring of the fitting results is fundamental to avoid local or nonphysical minima of the minimization process. Analogously, correlation indicators between each couple of optimized parameters should ideally be lower than 0.8 in absolute value and should never exceed 0.9. Eq. (5.5) underlines the importance of acquiring the EXAFS spectrum over the largest possible k-interval. Experimental data collected up to a high maximum k-value k_{max} allows to: (i) increase n_{ind}, as Δk increases; (ii) reduce the correlation between N_i and σ_i parameters; (iii) increase the ability to discriminate between two close distances. The distance resolution (Δr) of an EXAFS spectrum is indeed defined from k_{max} according to the relation:

$$\Delta r = \pi/(2k_{max}). \tag{5.6}$$

Two equally intense signals generated by the same scatterer located at r_1 and r_2 can indeed be singled out only in cases where the two oscillating functions $\sin(2k\, r_1)$ and $\sin(2k\, r_2)$ are able to generate at least a beat in the sampled k-range, and this occurs for $2k(r_1 - r_2) = \pi$. Consequently, in order to reach a distance resolution of $\Delta r = 0.1$ Å the EXAFS spectrum has to be collected up to about 16 Å^{-1}. The Δr parameter reported in Eq. (5.6) is the EXAFS-equivalent of the d-spacing parameter in diffraction experiments, obtained from the Bragg equation as $d_{min} = \lambda/[2\sin(\theta_{max})]$, and defines the accuracy in the reconstruction of the electron density.

5.2.4 Advantages and drawbacks of the technique

In a very concise way, we can summarize the main advantage of the XAFS technique as follows:

(i) The elemental selectivity allows to get information from the selected atomic species only.
(ii) It is applicable to very diluted species.
(iii) The relative high penetration depth of hard X-rays makes possible experiments on samples in contact with liquid or gases. This aspect is very important in catalysis, as *in situ* or *operando* experiments can be done.
(iv) The local nature of the X-ray absorption phenomenon, limited by the photoelectron mean free path λ, see Eq. (5.4), does not require long-range order, making liquids, gases and amorphous systems accessible to the investigation.

Using the same concise style, the main drawbacks of the technique can be summarized as follows:

(i) The laboratory X-ray sources using the Bremsstrahlung radiation emitted by rotating anodes provide a flux which is insufficient for most practical applications: synchrotron radiation sources are sorely needed.
(ii) The homogeneity of the sample is the main requisite for a proper application of the technique. If the dominant fraction (larger than 90 %) of the selected atomic species has the same local environment, then the coordination numbers and the bond-lengths extracted from the EXAFS data analysis are reliable. Conversely, when several phases are present, it is much more critical, although not impossible, to extract quantitative information from EXAFS. In such cases, the use of complementary information from independent techniques represents a fundamental approach to avoid misleading results. Examples on how to handle EXAFS data analyses when more phases are present are described in Sections 5.3.3, and 5.4.3. Alternatively, a way to overcome spatial heterogeneity of the sample using µm-focused X-ray beams is reported in Section 5.7.
(iii) The EXAFS investigation performed on relatively light elements embedded in matrices containing high Z elements is particularly critical, due to the larger absorption of the latter, see Figure 5.1b.

5.3 $CuCl_2/Al_2O_3$-based catalysts for ethylene oxychlorination

5.3.1 Industrial relevance of the $CuCl_2/Al_2O_3$ system

Nowadays, almost all the world production of polyvinyl chloride (PVC) is based on the polymerization of vinyl chloride, which is produced by the cracking of 1,2-dichloroethane (EDC), according to the following reaction:

$$C_2H_4Cl_2 \rightarrow C_2H_3Cl + HCl \tag{5.7}$$

$C_2H_4Cl_2$ is in turn obtained by two parallel processes, direct chlorination (5.8) and oxychlorination (5.9):

$$C_2H_4 + Cl_2 \rightarrow C_2H_4Cl_2 \tag{5.8}$$

$$C_2H_4 + 2HCl + \frac{1}{2}O_2 \rightarrow C_2H_4Cl_2 + H_2O \tag{5.9}$$

The latter reaction, recycling HCl produced by the cracking of EDC, is particularly important in industrial applications because it was specifically developed to reduce the consumption of Cl_2 and the exit of useless product (HCl) outside the cycle, in agreement with the modern requests of the chemical industry [63] that demand more environmental friendly reactions. Oxychlorination of ethylene offers a significant economic benefit in two ways: firstly, it uses cheap and abundant feedstock and, secondly, it incorporates the recycling of waste HCl produced in the cracking of EDC (5.7).

The oxychlorination reaction (5.9) is performed at 490–530 K and 5–6 atm using both air and oxygen in fluid or fixed-bed reactors. Commercial catalysts are produced by impregnation of γ-alumina with $CuCl_2$ (4–8 wt% Cu). Other metal-chlorides (mainly alkaline or alkaline earth chlorides) in a variable concentration are also added in order to improve the catalytic performances, making the catalyst more suitable for the use in industrial reactors. In particular, KCl is always present in the catalysts used in fixed-bed technologies, sometimes together with other alkali-metal chlorides, such as CsCl, NaCl or LiCl. Rare-earth-metal chlorides such as $LaCl_3$, added to $CuCl_2$ and KCl, are also claimed in the patent literature. $MgCl_2$ is the base additive in the catalysts used in fluid-bed processes, where alkali-metal (such as LiCl) or rare-earth-metal chlorides (such as $LaCl_3$) can also be added.

5.3.2 Preliminary *in situ* XAFS experiments

In the 2000–2003 period, Leofanti *et al.* achieved a significant improvement of the knowledge of the undoped $CuCl_2/Al_2O_3$ system [64–68]. In those studies several complementary techniques were adopted and, among them, EXAFS and XANES played a key role. Indeed, the $CuCl_2$ phase is so dispersed on the γ-Al_2O_3 support that even at a loading level as high as 9 wt% it is XRD silent [65]. This made EXAFS the only structural technique available for this system. XANES was widely used to determine the oxidation state of the active phase under the different catalyst conditions. In those studies it has been proved that two different copper species are present on the activated catalyst (dried at 50 K in N_2 flux): a surface copper aluminate, where the copper ions are hosted in the octahedral vacancies of γ-Al_2O_3, and a highly dispersed copper chloride phase. The former phase is the only one present at low Cu loadings (up to 0.95 wt% Cu per 100 m^2 support); the latter precipitates directly from the solution once the adsorptive capacity of alumina is exhausted [66–68].

5.3.2.1 The determination of the Cu-aluminate phase: How to avoid possible pitfalls in the EXAFS data analysis

When using Eq. (5.4) to fit the experimental EXAFS data, care must be taken to apply the correct structural model, *i.e.*, to hypothesize the correct atomic species surrounding the absorbing atom that contributes to the scattering process of the photoelectron. Indeed, the applied model defines the amplitude $F(k)$ and phase $\theta(k)$ functions used in the fit. A wrong choice of model will result in misleading results. The case of the Cu-aluminate phase mentioned so far is didactically emblematic to discuss this matter.

The Cu-aluminate phase was evidenced on a sample loaded with only 1.4 Cu wt% (hereafter Cu1.4) [67]. As the catalyst was prepared following the incipient wetness impregnation of a γ-Al_2O_3 with an aqueous solution of $CuCl_2$ $2H_2O$, two chemically reasonable possibilities can be suggested to model the first coordination shell around copper: (i) upon drying Cu^{2+} cations recombine with Cl^- anions forming a dispersed $CuCl_2$ phase; or (ii) upon drying Cu^{2+} occupies the cationic vacancies of the defective γ-Al_2O_3 support. Following model (i), Cu-Cl phases and amplitudes are adopted in Eq. (5.4) and the fit results in the bottom curves reported in Figure 5.2a. Adopting model (ii), Cu-O phases and amplitudes are adopted and the fit results in the top curves reported in Figure 5.2a. Model (ii) is preferred for the following reasons: (a) the better quality of the fit obtained with the Cu-O model (defined by the R- and χ^2-factors); (b) the reasonable values of the optimized parameters ($N = 4.8 \pm 0.5$; $r_{\text{Cu-O}} = 1.92 \pm 0.02$ Å: $\sigma = 8.9 \times 10^{-2}$ Å); (c) the result of the statistical F-test [69].

Comparing the two models, there is no doubt about which is the right one. But what would happen if only model (i) had been considered? What would have happened if model (ii) had not been taken into account from the beginning? Considering only model (i), could we understand that it is a wrong model? In this specific case the answer is yes, and we will report here two different methodological approaches that allow to discard model (i), regardless of the availability of an alternative model. The first approach consists in a simple comparison of the FT transform of the unknown material with those of model compounds, where the structural environment of the adsorbing atom is well established. Figure 5.2b reports the comparison between sample Cu1.4 and $CuCl_2$ $2H_2O$ model compound. The first coordination shells of Cu^{2+} in hydrated $CuCl_2$ exhibit two oxygens at 1.95 Å and two chlorines at 2.29 Å, responsible for the two peaks at 1.48 and 1.93 Å in the phase uncorrected FT of the EXAFS function reported as the green curve. As discussed in Section 5.2.3, without any phase correction, *i.e.*, performing the FT of the experimental $\chi(k)$ without considering the phase $2\delta(k) + \theta(k)$ terms inside the sine function in Eq. (5.4), the peaks of the radial distribution in the R-space

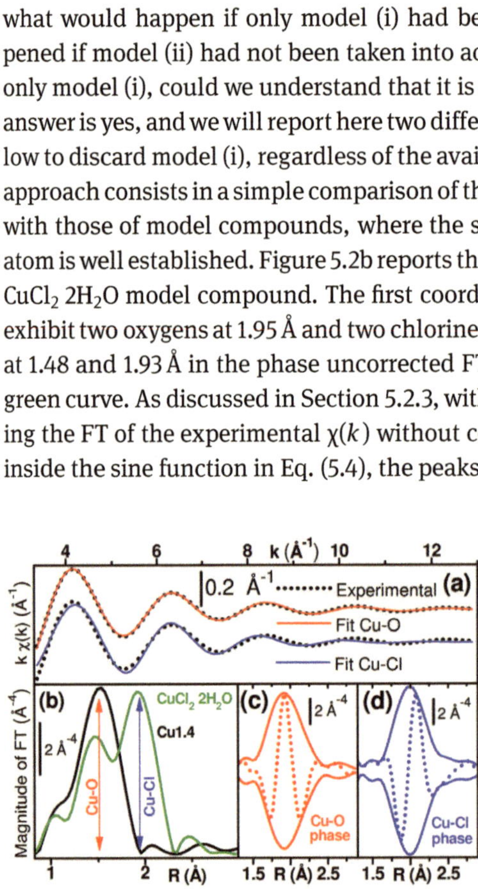

Fig. 5.2: Part (a): First shell filtered of the experimental EXAFS curve $k\chi(k)$, (black dotted lines) and comparison with best fit obtained using Cu-O (red curve) and Cu-Cl (blue curve) models. Part (b): k^3-weighted, phase-uncorrected, FT for the Cu1.4 sample and $CuCl_2$ $2H_2O$ model compound, black and green curves, respectively. Part (c): modulus (full line) and imaginary part (dotted line) of the k^3-weighted and Cu-O phase-corrected, FT for the Cu1.4 sample. Part (d): as part (c) using the Cu-Cl phase for the correction. Unpublished figure, reporting data published in [67].

always occur at shorter distances with respect to the real ones. Besides this technical aspect, it is evident that the radial distribution of sample Cu1.4 is not compatible with a Cu-Cl contribution, as it appears at a significantly shorter distance.

The second approach consists in the application of the Lee and Beni criterion [70]. This criterion affirms that when a given coordination shell of the absorbing atom A (the first in our case) is composed by only one atomic species B, located at the same distance (or with a very small spread in the distances), then the FT of the experimental $\chi(k)$ corrected by the $2\delta(k)+\theta(k)$ phase of the A–B couple will result in a radial distribution function centered at the right distance and having an imaginary part that is symmetric with respect to its modulus. Performing the FT of the $\chi(k)$ of sample Cu1.4 corrected by the Cu-Cl phases we obtain the radial distribution reported in Figure 5.2d. The fact that the imaginary part of that phase-corrected FT is not symmetrical with respect to its modulus automatically discards the model of a first shell of Cu^{2+} in Cu1.4 sample made by Cl atoms. Actually, it discards the fact that the first shell is made by only (or mainly) Cl atoms, but it cannot discard that some Cl atoms contribute (in a minority way to the first shell signal of the Cu1.4. sample). The definitive proof that in the first coordination of Cu^{2+} in Cu1.4 there are only O atoms, and thus that a Cu aluminate phase has been formed, is given by Figure 5.2d, showing that using Cu-O phases, the obtained FT has an imaginary part that is perfectly symmetrical with respect to its modulus.

5.3.2.2 Catalyst reactivity with the separate reactants: *In situ* XAFS experiments

Successively it has been shown that the active phase in ethylene oxychlorination reaction (5.9) is the highly dispersed $CuCl_2$ [68], following the three-step redox mechanism reported below [64, 65]:

$$2CuCl_2 + C_2H_4 \rightarrow C_2H_4Cl_2 + 2CuCl \qquad (5.10)$$

$$2CuCl + \frac{1}{2}O_2 \rightarrow Cu_2OCl_2 \qquad (5.11)$$

$$Cu_2OCl_2 + 2HCl \rightarrow 2CuCl_2 + H_2O \qquad (5.12)$$

This redox cycle has been determined by means of an *in situ* Cu K-edge XANES and EXAFS study, where the three reagents (C_2H_4, O_2 and HCl) have been separately dosed on the activated $CuCl_2/\gamma\text{-}Al_2O_3$ at 500 K in the given order (Fig. 5.3). In particular, the chlorination of ethylene, Eq. (5.10), occurs by reduction of $CuCl_2$ to CuCl, as testified by the red-shift of the Cu K-edge, see Figure 5.3a top curves. The oxidation of CuCl to an oxychloride, Eq. (5.11), is proved by both the opposite blue-shift of the Cu K-edge and the appearance of a low R-contribution in the |FT| of the EXAFS spectra, due to the insertion of oxygen in the first coordination sphere of Cu, see middle curves in Figure 5.3a and b, respectively. Finally, the catalytic cycle is closed by the re-chlorination of the oxychloride phase with HCl, Eq. (5.12), as demonstrated by the disappearance of the above mentioned oxygen contribution in the FT of the EXAFS spectrum, see red curve in the bottom of Figure 5.3b.

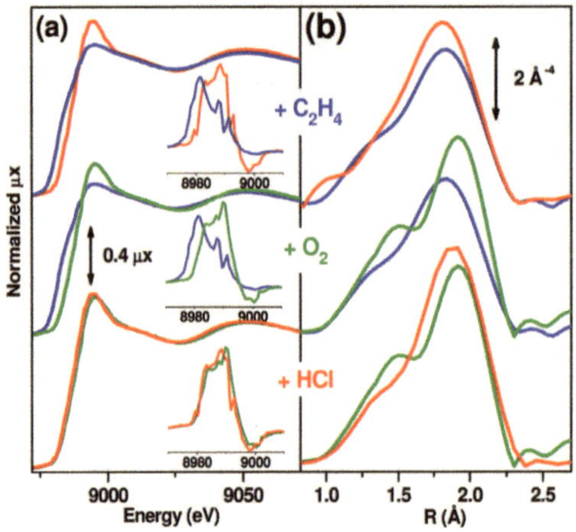

Fig. 5.3: Three-step redox mechanism of the ethylene oxychlorination reaction, Eq. (5.10)–(5.12) followed by in situ XANES and EXAFS, parts (a) and (b), respectively. From top to bottom the effect of the separate and successive dosage of the three reagents is reported: activated catalyst before (red line) and after (blue line) ethylene interaction; reduced catalyst before (blue line) and after (green line) interaction with O_2; oxidized catalyst before (green line) and after (red line) interaction with HCl. In all cases the interaction temperature was 500 K. Insets in part (a) report the derivative of the spectra reported in the main part. Unpublished figure reporting data collected at LURE DCI (EXAFS13 beam line) from [64, 65]; reproduced with permission from [13], copyright American Chemical Society 2013.

The spectra shown in Figure 5.3a demonstrate that the oxidized and the reduced form of the catalyst are easily recognizable by XANES spectroscopy. In particular, the XANES spectrum of the oxidized catalyst is characterized by an intense white line at 8995 eV, while that of the reduced catalyst shows a less intense white line at 8993 eV and exhibits a red shift of the edge of 5 eV. Even more informative are the derivative spectra (see insets), where the oxidized catalyst exhibits a strong maximum at 8988 eV and a negative peak at 8998 eV, while the reduced catalyst shows a maximum at 8981 eV.

More recently Muddada *et al.* [43, 44, 71, 72] extended the works of Leofanti *et al.* [64–68] from the bare to the LiCl-, KCl-, CsCl-, $MgCl_2$- and $LaCl_3$-doped $CuCl_2/Al_2O_3$ catalysts. Combining *in situ* XANES/EXAFS, IR spectroscopy of adsorbed CO, CO chemisorption and catalytic tests performed using a pulse reactor in depletive mode the authors succeeded in understanding the role of dopants (LiCl, KCl, CsCl, $MgCl_2$ $LaCl_3$) in determining the nature, relative fraction, reducibility and dispersion of Cu-phases on $CuCl_2/\gamma\text{-}Al_2O_3$ catalysts for oxychlorination reaction (*vide infra* the main parts of Figure 5.5 for the reduction by ethylene at 500 K, Eq. (5.10)). As already discussed above from the previous works, it was known that in the undoped catalyst two Cu phases coexist: Cu-aluminate and supported $CuCl_2$, the latter being the only active one [66]. In particular, the EXAFS and XANES data demonstrated the following trend in the ability of the dopant cation to compete with Cu^{2+} in the aluminate phase formation: $K^+ \leq Cs^+ \leq Li^+ \ll Mg^{2+} < La^{3+}$ [44]. In particular, a fraction close to 100 % of copper in the active phase is obtained by adding $MgCl_2$ or $LaCl_3$ dopants. Moreover, EXAFS directly, and IR indirectly, proved that the addition of KCl or CsCl (and less efficiently of LiCl) promotes the formation of mixed CuK_xCl_{2+x} or $CuCs_xCl_{2+x}$ phases, so altering the chemical nature of the active phase. XANES spectroscopy indicates that addition of $MgCl_2$ or $LaCl_3$ does not affect the reducibility by ethylene of the active $CuCl_2$ phase and that the reducibility of the new copper-dopant mixed chloride are in the order $CuCl_2 > CuLi_xCl_{2+x} > CuK_xCl_{2+x} > CuCs_xCl_{2+x}$ [44], (*vide infra* the main parts of Fig. 5.5).

5.3.3 Operando experiments and criteria used to face the presence of more than one phase in the sample

The determination of the rate determining step among reactions (5.10)–(5.12) has been successively achieved by a time-resolved XANES operando study [42, 43, 73]. The experiment was performed by feeding a cell containing a self-supported thin pellet of the catalyst with a diluted mixture of the three reagents ($C_2H_4 : HCl : O_2 : N_2 = 100 : 36.1 : 7.6 : 180$), representative of the fixed-bed process. In the course of the experiment the temperature was increased from 373 to 623 K and then decreased again to 373 K. Time-resolved XANES spectra were collected in dispersive geometry and the catalyst activity was simultaneously analyzed with a quadrupole mass spectrometer. The results of the experiment are summarized in Figure 5.4, part (a) for the heating step and part (b) for the cooling step of the experiment. The top curves in parts (a) and (b) report the evolution of the XANES spectra, while the bottom curves correspond to their first derivatives. Figure 5.4c summarizes the evolution of the Cu^{2+} fraction (red, orange and magenta curves) estimated from the XANES and compared with the catalyst activity reported in terms of oxygen consumption (scattered blue dots).

The XANES spectra reported in Figure 5.4a, and b indicate a progressive reduction of Cu^{2+} into Cu^+ during the heating step whereas almost no changes are observed in the XANES spectra collected in the cooling step. To evaluate the Cu^{2+} fraction for each spectrum along the heating or cooling step, two model spectra are needed, one for the fully oxidized catalyst and one for the fully reduced. The catalyst itself, after activation in N_2 flux up to 500 K before and after reduction in C_2H_4 atmosphere at 500 K, has been used for this purpose. Three independent methods have been used to quantify the fraction of Cu^{2+} belonging to the active phase on the catalyst in the i-th spectrum, x_i. The first method concerns determining the energy (E_i) position of the edge, arbitrarily defined as the energy corresponding to normalized $\mu x = 0.5$:

$$x_i = (E_i - E_{Red})/(E_{Ox} - E_{Red}), \tag{5.13}$$

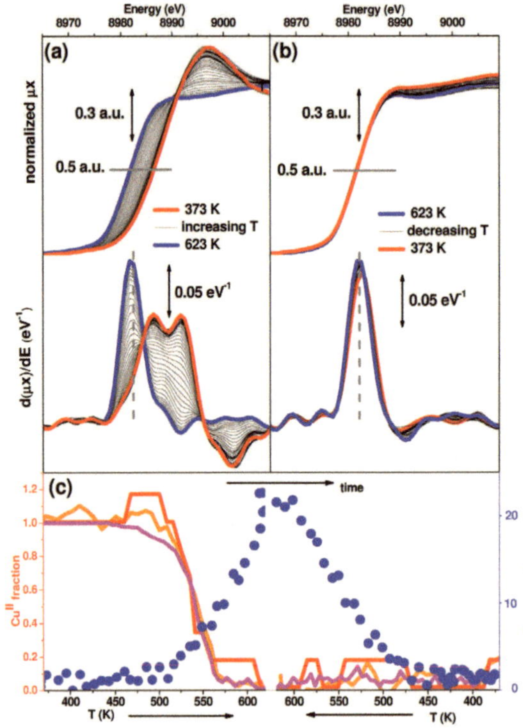

Fig. 5.4: Part (a): XANES spectra (top) and corresponding derivatives (bottom) of $CuCl_2/Al_2O_3$ system in ethylene oxychlorination environment, during the heating step of the experiment (from 373 K, blue curve, to 623 K, red curve, by 12 K min^{-1}). Part (b) as part (a) for the cooling step of the experiment (from 623 K, red curve, to 373 K, blue curve, by 12 K min^{-1}). XANES spectra have been collected each 30 s, corresponding to a ΔT of 6 K. Part (c) summarizes the results of the operando experiment (both heating and cooling steps, left and right parts, respectively), by correlating the catalyst activity, monitored as O_2 consumption (dotted scattered blue points, right ordinate axis) with the fraction of Cu^{2+} (left ordinate axis) red, orange and magenta lines, evaluated according to methods reported in Eqs. (5.13), (5.14) and (5.15)-(5.16), respectively. Figure reporting data collected at ESRF ID24 beamline and published in [42, 73]; reproduced with permission from [13], copyright American Chemical Society 2013.

where E_{Ox} and E_{Red} are the energy of the edge in the oxidized and reduced reference spectra respectively; and E_i the position of the edge in the i-th spectrum. Cu^{2+} fractions evaluated from this method are reported as red lines in Figure 5.4c.
The second method uses the intensity (I) of the first maximum due to the Cu^{+} component in the derivative spectra at 8982 eV:

$$x_i = 1 - I_i/I_{Red}, \tag{5.14}$$

where I_{Red} refers to the spectrum of the reduced reference. Cu^{2+} fractions evaluated from this method are reported as orange lines in Figure 5.4c. The peak at 8981 eV has been chosen because it is the most intense feature of the derivative XANES spectra of Cu^{+} species and it falls in a region where the Cu^{2+} species have a null derivative. This method cannot be translated to directly measure the fraction of Cu^{2+} using the most intense Cu^{2+} features at 8985 and 8992 eV, because in this region the derivative of Cu^{+} species is still strong.

The last method consists in the simulation of the i-th spectrum as the linear combination of the two reference spectra according to Eq. (5.15) [31, 43, 45, 68, 73]:

$$Xanes_i^{Theo}(E_j, x_i) = x_1 Xanes^{Ox}(E_j) + (1 - x_i) Xanes^{Red}(E_j) \tag{5.15}$$

The fraction of oxidized copper species, x_i, is optimized by a least-squares method, minimizing for each i-th spectrum the $\mathbf{F}(x_i)$ function given by:

$$\mathbf{F}(x_i) = \sum_{j=1}^{N} \left[Xanes_i^{Exp}(E_j) - Xanes^{Theo}(E_j, x_i)\right]^2 \tag{5.16}$$

where E_j ($j = 1, 2, \ldots N$) are the energies where the experimental spectra have been sampled. Cu^{2+} fractions evaluated from this method are reported as magenta lines in Figure 5.4c. Although this last method is more time-demanding that the previous ones, it is also more accurate, as it accounts for all the points of the spectrum.

Comparing in Figure 5.4c (left axis) the fraction of Cu^{2+} species (x_i) evaluated according to methods (5.13), (5.14) and (5.15)–(5.16) for the series of spectra reported in Figure 5.4a,b we can conclude that in all cases the agreement among the three methods is rather good, as they provide values in relative agreement within $\pm 5\,\%$. It is interesting to note that the x_i fraction evaluated according to method (5.13) has a step-like trend, because of the discreteness of the energy scale measurable with a position-sensitive detector.

From the whole set of data summarized in Figure 5.4c the following can be commented. At the starting point (373 K) only Cu^{2+} is present and the catalyst is inactive. Both O_2 conversion and Cu^{2+} reduction start in the same temperature range (470–490 K) and progressively increase upon increasing the temperature. Cu^{2+} reduction is completed at 600 K (Fig. 5.4c). During the cooling step, O_2 conversion progressively declines becoming negligible in the 490–470 K range, while the oxidation state of Cu does not change anymore. These results provide evidence that at the typical oxychlorination temperature, Cu^{+} is the dominant phase and the rate determining step is the oxidation of CuCl, according to Eq. (5.11). In order to understand the low efficiency of the oxidation process, the catalyst at the end of the cooling step has been subjected to two different oxidizing treatments at increasing temperatures (373–623 K): the former with diluted O_2, the latter with a diluted O_2-HCl mixture. The first treatment causes the complete Cu oxidation already at 373 K, while the second leaves the Cu^{+} unchanged up to 550 K, and, even at 623 K, Cu^{+} is still present. The authors concluded that HCl acts as a poison for Cu^{+} oxidation and is responsible for the prevailing reduced state of copper during reaction [42, 43].

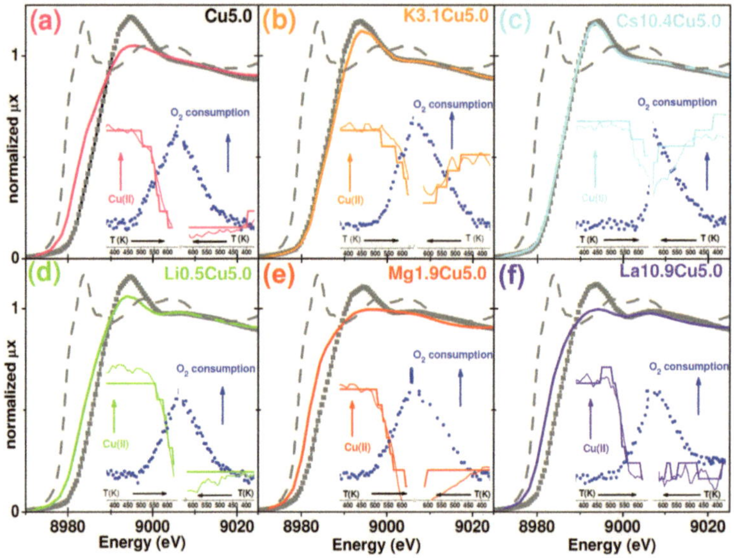

Fig. 5.5: Main parts: XANES static spectra (LURE DCI, EXAFS13 beam line) of doped samples before (grey scattered curves) and after (colored solid lines) reduction by ethylene at 500 K compared with a spectrum of pure Cu(I) chloride (dashed grey line; same curve in all frames). The insets show the results of operando experiments (ESRF ID24 beamline) reporting the evolution of the Cu(II) fraction along the heating and cooling runs estimated from the edge shift (Eq. (5.13) bold colored line) and from the intensity of the maximum of the first derivative (Eq. (5.14) full colored line) together with the catalyst activity (scattered blue dots). Figure reporting data published in [42, 43, 72]; reproduced with permission from [13], copyright American Chemical Society 2013.

The key role of potassium chloride dopant, present in the industrial catalysts used in fixed-bed technologies, has been highlighted in a similar experiment [42]. It has been shown that the $KCl/CuCl_2/\gamma\text{-}Al_2O_3$ catalyst behaves differently from the base one, working in a prevailingly oxidized state (Fig. 5.5b). Combining operando XANES experiments with catalytic tests of ethylene conversion in pulse reactors and with IR experiments of adsorbed CO, it was concluded that the active phase of the $KCl/CuCl_2/\gamma\text{-}Al_2O_3$ is a mixed chloride (K_xCuCl_{2+x}) phase [42], which reduces the ability of the active surface to adsorb ethylene and/or transfer two Cl atoms to each ethylene molecule. Although not detectable by XRD owing to too small crystal size [66], the formation of the mixed-chloride double compound was suggested by IR spectroscopy of adsorbed CO [42] and proved by EXAFS [44].

More recently, Muddada *et al.* [43] extended the works of Lamberti *et al.* [42, 73] to the LiCl-, CsCl-, $MgCl_2$-, $CeCl_4$- and $LaCl_3$-doped $CuCl_2/Al_2O_3$ catalyst. Some of these results are summarized in Figure 5.5. The authors coupled the catalyst activity (monitored with a pulse reactor working in both non-depletive and depletive modes) with time-resolved XANES spectroscopy performed under operando conditions (as done for the undoped catalyst, see Figs. 5.3 and 5.4); they succeeded in understanding, at the atomic level, the role played by additives in the chemistry of the industrial catalyst used in the ethylene oxychlorination. Both KCl (Fig. 5.5b), and CsCl (Fig. 5.5c), form in reaction conditions a mixed phase with $CuCl_2$, which strongly modify the catalyst behavior [43]. In particular, these additives were able to reverse the rate determining step, from the CuCl oxidation (typical of the undoped catalyst) to the $CuCl_2$ reduction *i.e.*, from Eq. (5.11) to Eq. (5.10). The change in the rate determining step is a consequence of the decrease of the rate of the latter reaction, thus of the overall activity of the system. For all remaining additives (Fig. 5d–f) the rate determining step remains the CuCl oxidation, as for the undoped catalyst [43], because the additive chlorides do not mix with $CuCl_2$, which remains the active phase.

5.4 Structural and electronic configuration of Cp_2Cr molecules encapsulated in PS and Na-Y zeolite and their reactivity towards CO

In the last decades much effort has been devoted to find new methods for metal incorporation and catalyst immobilization inside nanoporous scaffolds [74]. Inside this wide field of research, the inclusion of organometallic complexes inside host frameworks gives the possibility of disparate applications such as electronic devices, optical materials or heterogeneous catalysis [75, 76] gaining even more interest.

Metallocenes are organometallic substrates used to investigate potential reactivity enhancement; among them, chromocene (Cp_2Cr) is a challenging candidate both for applicative reasons, being the precursor of the well-known Union Carbide olefin catalyst [77], and for theoretical causes since there is still a challenging problem in modeling open-shell kinds of molecules.

Not only has theory alone not been able to solve all the open questions on this topic, but even the experimental approaches adopted so far to investigate properties and reactivity of these systems (such as matrix isolation technique [78] and dissolution in a solvent [79, 80]) suffered from some disadvantages and the results were not in agreement with *ab initio* calculations. The problems consisted in the air sensitivity of the molecules of Cp_2Cr (that necessitates working in a controlled atmosphere) and their tendency to form clusters, so that most of the techniques give information about the clusters and not about the isolated molecules or about the active species useful for the catalysis.

5.4.1 Structure of Cp_2Cr encapsulated in PS and Na-Y zeolite matrices

Estephane et al. [37, 81–83] developed new strategies to obtain spectroscopic information on isolated molecules and on their reactivity, in particular towards CO. Two matrices were adopted as "scaffolds" to encapsulate and isolate molecules in their cavities:

- A non-polar nanoporous polystyrene (PS) matrix (average dimensions of pores 100 Å, high surface area $\approx 1000\,m^2g^{-1}$) that can act as a "solid solvent" molecularly dispersing the chromocene inside its nanopores (Fig. 5.6a).
- A polar Na-Y zeolite (Fig. 5.6b), that is expected to enhance the reactivity of the hosted molecules by means of the internal electrostatic fields. In fact, zeolitic voids could function as "nanoscale reaction chambers", where the reactivity of guest organometallic complexes can provide molecular insights into the elementary steps of heterogeneous catalysis.

A systematic XAFS and IR study allowed the authors to obtain a complete structural, electronic and vibrational insight into the isolated molecule and its interaction products with CO in the two different environments [37, 81–83]. In these experiments CO sent from gas phase was able to easily diffuse in the porous matrices, overcoming the diffusion problems of experiments performed dispersing Cp_2Cr in toluene [79, 80].

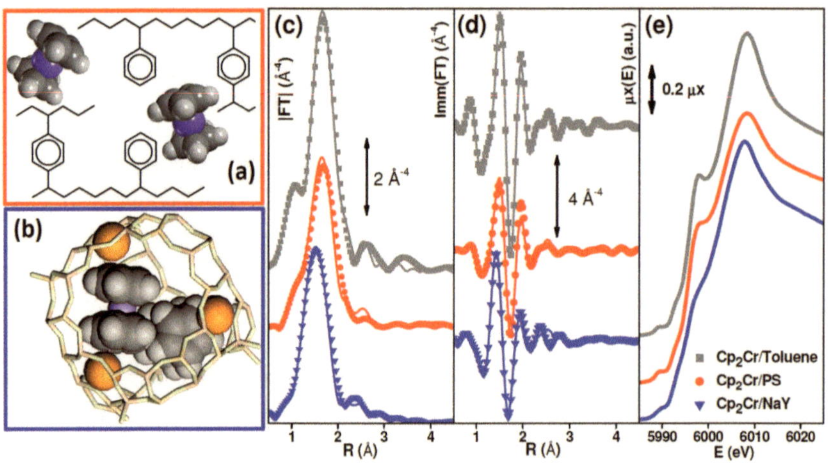

Fig. 5.6: Part (a): Two Cp_2Cr molecules, reported as Corey-Pauling-Koltun (CPK) spheres: violet, dark grey, and light grey for Cr, C and H atoms respectively, encapsulated inside a 2D portion of PS matrix. Part (b) as part (a) where the Cp_2Cr molecules are encapsulated in a 3D portion of a zeolite cell (supercage cavity [84]); framework atoms are represented in the stick mode (Si and Al, light orange; O, light yellow) and Na^+ cations as orange CPK spheres. Part (c): modulus (a) of the k^3-weighted, phase-uncorrected, FT of EXAFS signals for Cp_2Cr/toluene (top grey curves), Cp_2Cr/PS (middle black curves), and Cp_2Cr/Na-Y (bottom blue curves experimental data as square dots and best fits as full lines). Part (d) as part (c) for the imaginary parts. Part (e) corresponding XANES spectra. Unpublished figure reporting data published in [37].

For both matrices, the absence of clustered Cp_2Cr molecules was determined by XAFS measurements. Results were compared with those obtained on Cp_2Cr in toluene solution (assumed as a model for isolated molecules). From the k^3-weighted, phase-uncorrected Fourier transform (FT) of the EXAFS signals reported in Figure 5.6c,d (modulus and imaginary part) it is clearly evident that they are very similar. In fact they present a first-shell contribution around 1.6 Å due to 10 Cr-C SS paths and some MS contributions at higher distances but they do not show any other peak assignable to a Cr-Cr path characteristic for clusters of chromocene molecules. Also the three XANES spectra (see Section 5.4.2 for a more detailed description) exhibit basically the same features confirming that the molecules are really isolated (Fig. 5.6e). However, if we concentrate on the numerical results of the best fit for the three systems (Tab. 5.1), in the case of Cp_2Cr/Na-Y the molecule seems to be slightly distorted and in fact we obtain a shorter mean distance $\langle r_{Cr\text{-}C} \rangle$ resulting in a shift to shorter R values in FT of the EXAFS signal. The higher value of σ^2 Debye–Waller factor and the features less defined in the corresponding XANES spectrum indicate a higher heterogeneity of the species. This is assigned to different local electric fields experienced by Cp_2Cr molecules due to different statistical cation distribution in the supercages [84], see Figure 5.6b.

Tab. 5.1: Summary of the structural parameters optimized in the fit of the EXAFS data for the Cp_2Cr/toluene, Cp_2Cr/PS and Cp_2Cr/Na-Y systems. The fits run in R-space, in the 1.0–4.5 Å range. Single S_0^2 and ΔE_0 have been optimized for all SS and MS paths. Unpublished Table, reporting data fron Refs. [37, 81, 83].

	Cp_2Cr/toluene	Cp_2Cr/PS	Cp_2Cr/Na-Y
S_0^2	0.7 ± 0.1	0.7	0.9 ± 0.1
ΔE_0 (eV)	0 ± 1	$+1 \pm 1$	-4 ± 2
$\langle r_{Cr\text{-}C} \rangle$ (Å)	2.169 ± 0.006 (DFT 2.201)	2.165 ± 0.008	2.11 ± 0.01
$\sigma^2_{Cr\text{-}Cp}$ (Å^2)	0.003 ± 0.001	0.0070 ± 0.0005	0.010 ± 0.002
n° variables	4	3	4
R_{factor}	0.020	0.037	0.057

5.4.2 Determination of the electronic structure of Cp_2Cr by combined UV-Vis and XANES spectroscopies

In the same work, Estephane *et al.* [37] also reported a didactically instructive approach to determine the electronic structure of the Cp_2Cr molecule based on the combination of UV-Vis and XANES spectroscopies.

When Cp_2Cr is sublimed onto the Na-Y powder, the white zeolite acquires a light salmon color. As a consequence, the UV-Vis spectrum of the Cp_2Cr/Na-Y system (Fig. 5.7a, gray curve) is characterized by three main components centered around 21,000, 30,000 (responsible for the color), and 43,000 cm^{-1} (henceforth labeled as components **I**, **II**, and **III**, respectively). These bands are very similar to those observed in the spectrum of chromocene in solution [85, 86]. A simplified scheme of the electronic structure of Cp_2Cr is reported in Figure 5.7b. In the idealized D_{5d} symmetry, the

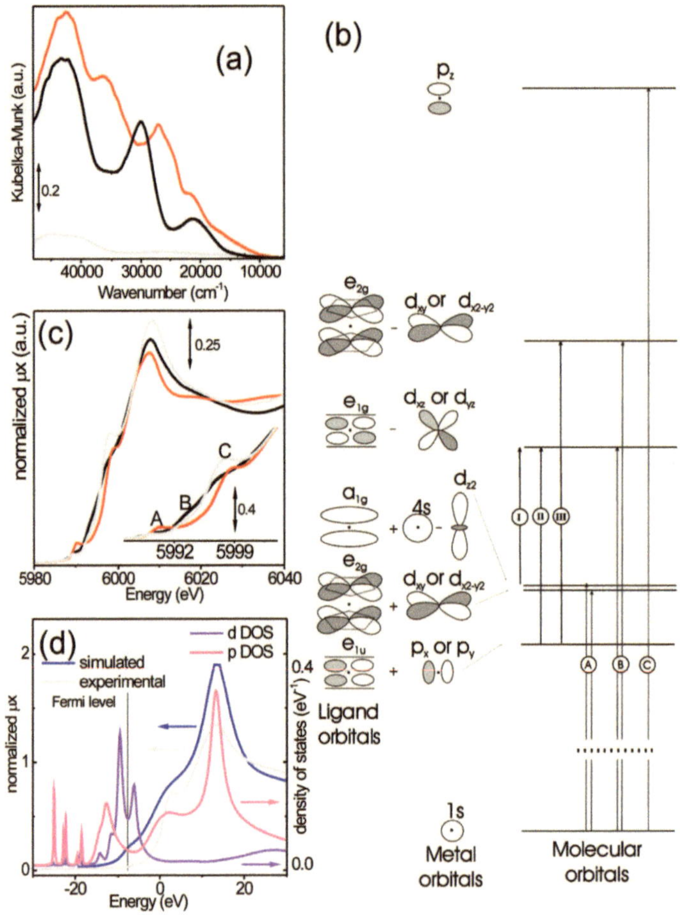

Fig. 5.7: Part (a): UV-Vis DRS spectra of Cp_2Cr/Na-Y before (black line) and after (red line) CO dosage (equilibrium pressure P_{CO} = 150 Torr). Also the spectrum of Na-Y zeolite (gray line) is reported for comparison. Part (b): rough representation of the electronic transitions and the molecular orbitals involved in the electronic transitions responsible for the UV-Vis and XANES spectra. Part (c): XANES spectra of Cp_2Cr/Na-Y before (black line) and after (red line) CO dosage (equilibrium pressure P_{CO} = 150 Torr). Also the spectrum of Cp_2Cr in toluene is reported for comparison (light gray). The inset reports a magnification of the pre-edge region. Part (d), left ordinate axis: simulated XANES spectrum of Cp_2Cr (blue) compared to experimental one for Cp_2Cr in toluene (light gray, horizontally translated to match the white line position). Right ordinate axis (as indicated by the arrows) shows the *p* and *d* projected DOS for the Cr atom (magenta and violet, respectively). Reproduced with permission from [37], copyright American Chemical Society (2009).

10 π orbitals of the two Cp rings form the symmetry-adapted combinations a_{1g}, a_{2u}, e_{1g}, e_{1u}, e_{2g}, and e_{2u}. Interaction of these ligand-centered orbitals with the metal $3d$, $4s$, and $4p$ valence atomic orbitals of appropriate symmetry generates the molecular orbitals shown in the right-hand side of the diagram.

The Cp_2Cr ground state is $^3E_{2g}$, resulting from the $e_{2g}^3\ a_{1g}^l$ configuration [85]. The transitions of one of the 16 valence electrons into a non-occupied molecular orbital are responsible for the observed optical spectrum (Fig. 5.7a). Only the lowest frequency component (21,000 cm^{-1}) has been definitely assigned in literature in terms of a *d*-*d* transition ($^3E_{2g} \rightarrow {}^3E_{1g}, {}^3E_{2g}, {}^3A_{1g}, {}^3A_{2g}$). Based on this, band **I** is due to an electronic transition from the HOMO to the LUMO [87–89]. The *d*-*d* character of transition **I** is easily explained by considering that both the HOMO and the LUMO are primarily metal in character (d_{xy,x^2-y^2} for the HOMO and $d_{xz,yz}$ for the LUMO) [90]. The bands at 30,000 and 43,000 cm^{-1} were assigned to ligand-to-metal charge transfer transitions [37]. The key feature from this investigation is that the UV-Vis spectrum of Cp_2Cr/Na-Y clearly indicates that Cp_2Cr is simply encapsulated as a neutral Cp_2Cr molecule inside the supercages of Na-Y zeolite and no further reaction occurs, since similar spectra have already been reported for chromocene in solution [85, 86] and in a frozen argon matrix [91].

Estephane *et al.* [37] used the same simplified scheme also to understand the Cr K-edge normalized XANES spectrum of Cp_2Cr, in the approximation that the core hole generated in the Cr(1*s*) orbital by the X-ray absorption does not strongly perturb the molecular levels. The XANES spectrum of Cp_2Cr in Na-Y zeolite (black curve in Fig. 5.7c) is compared to that of Cp_2Cr in toluene solution (solid gray curve).

The spectrum of the reference sample is characterized by three main features in the pre-edge region at 5989.2, 5992.8, and 5997.7 eV (labeled **A**, **B**, and **C**, respectively), followed by an intense and sharp white line centered at 6008.2 eV. To allow an assignment of these features described above, the authors performed a simulation of the XANES spectrum of the Cp_2Cr molecule with FEFF8.4 code [92, 93]. Figure 5.7d compares the simulated XANES spectrum of the isolated molecule (blue) with the experimental one collected in toluene solution (light gray), properly translated to match the energy position of the white-line transition. The simulated spectrum is in good agreement with the experimental one and is able to reproduce the energy position and the relative intensity of all the features. In Figure 5.7d (right ordinate axis) the calculated projected DOS for the Cr atom is also reported on the same energy scale. Comparison of the simulated spectrum with the projected DOS allows the assignment of the main XANES features, which can be divided in two families. At lower energy, features **A** and **B** are due to the transition between 1*s* orbital and Cr levels characterized by a significant *p*/*d* hybridization. At higher energy, C and white-line features are due to the transition of a 1*s* electron into almost "pure" *p* levels. On the basis of this simulation, the intense **C** feature is assigned to the dipole-allowed $1s \rightarrow 4p_z$ transition (see the electronic scheme reported in Fig. 5.7b) and is consequently very sensitive to coordination and geometry variation along the z axis of the molecule [94]. For a complete assignment of features **A** and **B**, a combination of XANES (Fig. 5.7c) and UV-Vis (Fig. 5.7a) spectra is mandatory. The energy difference between the levels originated from the $e_{1g} - d_{xz}, d_{yz}$ (LUMO) and $e_{2g} - d_{xy}, d_{x2-y2}$ anti-bonding combinations is 13,000 cm^{-1} (1.6 eV). This value, obtained from the UV-Vis spectrum (frequency difference between band **III** and band **II**), is too narrow to be resolved in the XANES spectrum owing to the core-hole broadening. The shoulder **B** is consequently assigned to the unresolved electronic transitions from 1*s* to these levels. Finally, the **A** feature, found 3.6 eV (29000 cm^{-1}) below **B**, is assigned to the dipole-forbidden transitions from the 1*s* to the HOMO levels. In fact, the UV-Vis spectrum reveals that the almost degenerated and partially unoccupied HOMO levels lie 21,000 cm^{-1} below the LUMO level and 34,000 cm^{-1} below the levels originated from the $e_{2g} - d_{xy}, d_{x2-y2}$ antibonding combinations, respectively (average value of 27,500 cm^{-1}, as XANES does not resolve the two final states).

The XANES spectrum of the Cp_2Cr/Na-Y system presents basically the same main features discussed so far for Cp_2Cr in toluene, corroborating the conclusion of the UV-Vis study, *i.e.*, that Cp_2Cr is confined in the Na-Y supercages without changes of the Cr oxidation state. However, the spectrum is less resolved, indicating a higher heterogeneity of species. This is assigned to different local electric fields experienced by Cp_2Cr molecules, due to different statistical cation distribution in the supercages of the Y zeolite. In further detail, feature **A** remains basically unchanged, feature **B** increases in intensity, while an inverse behavior is observed for both feature **C** and for the white line. Since the most affected feature is the **C** one, XANES data evidenced an appreciable structural modification of the Cp_2Cr molecule along the z axis, due to the interaction with the zeolitic walls and to the presence of strong local electric fields [95].

5.4.3 Reactivity of Cp_2Cr hosted in PS and in Na-Y zeolite towards CO: IR and XAFS results

Estephane *et al.* [37, 81] followed the interaction of CO with both the Cp_2Cr/PS and Cp_2Cr/Na-Y matrix systems with IR spectroscopy. In the case of PS, the formation of a time-stable $Cp_2Cr \cdots CO$ monocarbonyl complex was demonstrated, in fact upon dosing CO at RT on the Cp_2Cr/PS system, an intense band at 1900 cm^{-1} appears in the FTIR spectrum (Fig. 5.8a), assigned to the ν(CO) of the $Cp_2Cr(CO)$ complex. This band was previously observed in literature for direct addition of CO to a toluene solution but the complex formed was labile [80], so the improvement by working with a "solid solvent" matrix is clearly evident. In a successive work, Estephane *et al.* [83] used the EXAFS spectrum of the $Cp_2Cr(CO)$ complex formed inside the PS matrix to validate the *ab initio* models predicting a triplet → singlet spin transition in the $Cp_2Cr(CO) + CO \rightarrow Cp_2Cr(CO)$ reaction.

The case of Na-Y is more complex as it presents a rapid evolution in time. Upon CO addition the Cp_2Cr/Na-Y system an intense band in the ν(CO) region at 1850 cm^{-1} immediately appears in the FTIR spectrum (bold blue curve in Fig. 5.8b), and can be assigned to a monocarbonyl complex of the type $Cp_2Cr(CO)$. The red-shift of the value with respect to PS was explained by a high ionic environment [37]. By waiting in time this band is progressively eroded and three bands at 1893, 1773, and 1724 cm^{-1} (red

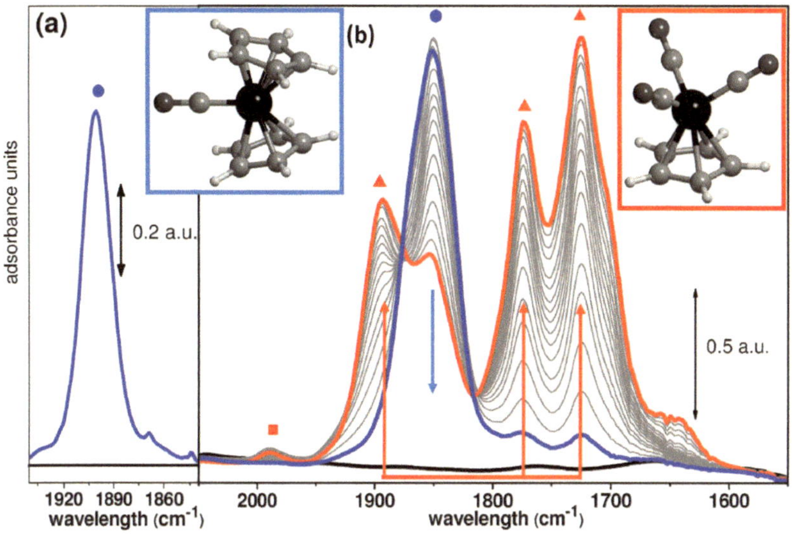

Fig. 5.8: Part (a): FTIR spectra collected at RT on the Cp_2Cr/PS system before (black) and after (blue) dosage of CO. No time evolution is observed in this experiment. Part (b): FTIR spectra collected at RT on the Cp_2Cr/Na-Y system before (black), immediately after dosage of CO (blue), after 1 h (red) and intermediate times (grey curves). The blue inset reports the structure of the $Cp_2Cr(CO)$ complex responsible of the IR band at 1900 cm^{-1} (PS) or at 1850 cm^{-1} (Na-Y). The red inset reports the structure of the $[CpCr(CO)_3]$, complex responsible of the IR triplet at 1893, 1773 and 1724 cm^{-1}. Unpublished figure reporting data published in [37, 81].

triangles in Fig. 5.8b) simultaneously appear accompanied by a weak component at $\sim 1980\,cm^{-1}$ (red square) growing in a similar way. The evolution was interpreted in terms of an evolution of the monocarbonyl precursor into a tricarbonyl charged species $[CpCr(CO)_3]^-$, after the loss of a Cp ring. This transformation is favored by the intense electric field present inside the zeolite's cavities. In the same way, the weak component at $\sim 1980\,cm^{-1}$ was assigned to a fraction of $[Cp_2Cr(CO)]^+$ species [37].

The k^3-weighted FT of the EXAFS data and XANES spectra (Fig. 5.9a,b) reflect the behavior observed by IR and highlight structural variations undergone by the Cp_2Cr molecule upon CO addition. In particular, for the Cp_2Cr/PS system, the modulus (top spectra in Figure 5.9a) exhibits an increase of the intensity and a shift toward higher R values of the first shell, and together with the numerical results of the good fit (made starting from *ab initio* models of a monocarbonyl species) prove the insertion of a CO ligand and a consequent rearrangement of the Cp rings with elongation of the average $\langle r_{Cr\text{-}C} \rangle$. An additional high R component appears also, centered around 2.5 Å but with different phase with respect to Cr-Cp MS (see imaginary part Fig. 5.9a), that is ascribed to the strong collinear Cr-C-O MS contributions. In a similar way, the modification of all the pre-edge XANES features confirms that the Cp_2Cr molecules lose their symmetry and undergo a relevant distortion after interaction with CO, see Fig. 5.9b.

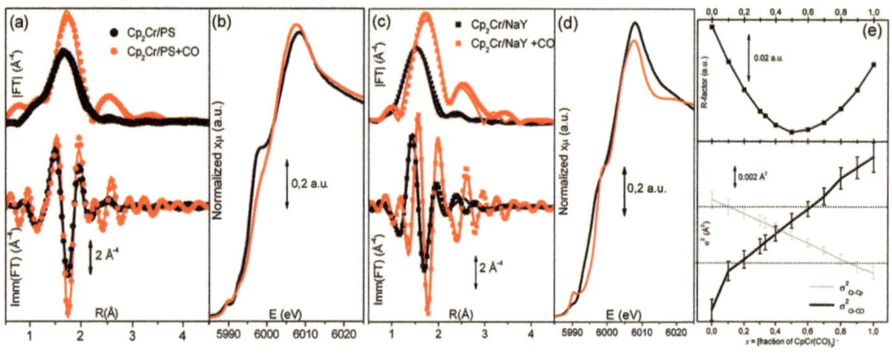

Fig. 5.9: Part (a): Modulus (top) and imaginary part (bottom) of the k^3-weighted, phase uncorrected, FT of the EXAFS spectra of the Cp_2Cr/PS systems before (black curves) and after (red curves) CO dosage. The experimental data (scattered) are superimposed with the best fit (full lines). Part (b): as part (a) for the XANES spectra. Parts (c) and (d): as parts (a) and (b) for the Cp_2Cr/Na-Y system. Part (e): results of the two phases fit on Cp_2Cr/Na-Y + CO system trend of R-factor and of Debye–Waller factors in function of the relative percentage of products of reaction. Unpublished figure reporting data published in [37, 82, 83].

In the Cp_2Cr/Na-Y system, XANES spectrum upon CO addition shows a strongly modified white-line intensity and better resolved pre-edge features, suggesting a better molecular definition of the Cr species formed upon CO contact. The $[CpCr(CO)_3]^-$ and $[Cp_2Cr(CO)]^+$ product species hypothesized from the IR study could justify this behavior even if the position of the edge is almost unaltered and the two products should be characterized by a different formal oxidation state, but it is possible that the charges are delocalized across the entire molecule and not localized on the Cr center.

Combining indications obtained by previous techniques with quantum mechanical calculations we simulated the EXAFS signal of the possible products of reaction between Cp_2Cr/NaY and CO. Because the time of CO contact in the XAS experiment is much longer than that of IR, the initial species Cp_2 Cr(CO) is not expected to be detected by this technique, while only the final state with the copresence of the $[CpCr(CO)_3]^-$ and $[Cp_2Cr\text{-}(CO)]^+$ species is expected to be observed. We initially performed two separate fits by supposing the presence of only one of the two species but both of them gave physically unreliable numerical results (underlined values in Table 5.2). Therefore we tried to quantitatively evaluate their relative amounts by means of a two-phase fit.

A phase fraction parameter x has been added to the fit, weighting the signals of the species $[CpCr(CO)_3]^-$ and $[Cp_2Cr(CO)]^+$ by x and $(1-x)$, respectively. Other parameters have been constrained in order to keep fixed the same number of variables optimized to 6, as for the two previous fits, so that they are directly comparable. The good quality of the fit is further appreciable in Figure 5.9c, and looking to the results, the R_{factor} is significantly improved and, more importantly, enforcing the x value to move from 0 to 1 with 0.1 steps all optimized parameters lie within the physical acceptable ranges in

Tab. 5.2: Summary of the structural parameters optimized in the fitting of the EXAFS data for the Cp_2Cr/Na-Y + CO systems. The fits were performed in R-space in the 1.0–4.5 Å range. Not optimized parameters are recognizable by the absence of the corresponding error bars. Underlined parameters evidence mathematical values of the fit that are physically not acceptable. The 2-phases fit resulted in an optimized value of 0.50 with an error of 0.07. Unpublished table reporting data from Ref. [37].

	1 phase $[CpCr(CO)_3]^-$	1 phase $[Cp_2Cr(CO)]^+$	2 phases $[CpCr(CO)_3]^-$	$[Cp_2Cr(CO)]^+$
S_0^2	0.9 ± 0.2	<u>0.6 ± 0.1</u>	$0.9{*}x$	$0.9{*}(1-x)$
ΔE_0 (eV)	1 ± 2	1 ± 2	3 ± 1	
$\langle r_{Cr\text{-}C} \rangle$ (Å)	2.193 ± 0.009	2.20 ± 0.01	2.227 ± 0.006	2.196 ± 0.006
$\sigma^2_{Cr\text{-}Cp}$ (Å^2)	<u>0.001 ± 0.001</u>	0.004 ± 0.002	0.0047 ± 0.0006	
$\langle r_{Cr\text{-}CO} \rangle$ (Å)	1.82 ± 0.02	1.85 ± 0.01	1.822 ± 0.006	1.893 ± 0.006
$\sigma^2_{Cr\text{-}CO}$ (Å^2)	<u>0.012 ± 0.002</u>	<u>0.000 ± 0.001</u>	0.006 ± 0.001	
n° variables	6	6	6	
R_{factor}	0.046	0.035	0.013	

the 0.3–0.6 interval. Since the phase fraction parameter x is optimized to 0.50 ± 0.07, as confirmed by the minimum in the R-factor trend in the function of x in Figure 5.9e, Estephane *et al.* concluded that the two species are almost equi-populated [37].

Summarizing, PS and Na-Y zeolite matrices can both act as "solid solvent" for the chromocene molecules, thus permitting the study of their interactions with simple reagents sent into the matrices in gas phase (an operation that is much more difficult in the case of a liquid solution). The differences between the two matrices are reflected in the molecules encaged, slightly in terms of geometry but especially in terms of their reactivity. In fact, interaction of Cp_2Cr with CO in PS comports the addition of only one CO, while in the zeolite a higher, but also much more complex, reactivity is observed giving as product new charged complexes not only monocarbonyls but also $[CpCr(CO)_3]^-$ where one Cp is substituted by three CO molecules.

5.5 Transition metal complexes in solution: The *cis*-$[Ru(bpy)_2(py)_2]^{2+}$ case study

The structure of a transition metal complex is influenced by its chemical environment, so some differences are expected moving from the crystalline phase to the dispersed phase either in different solvents or after grafting on different high surface area supports. Since relevant applications require dispersed transition metal complexes, the knowledge of their structures can be very important. Diffraction techniques cannot be applied in such conditions, and theoretical and/or spectroscopic approaches have to be considered. In this respect, EXAFS is one of the most suitable spectroscopic approaches. In particular, in this section we will discuss the case of the *cis*-$[Ru(bpy)_2(py)_2]^{2+}$ complex in aqueous solution, investigated using EXAFS spectroscopy [96].

From a methodological background, the procedure hereinafter reported for *cis*-$[Ru(bpy)_2(py)_2]^{2+}$ is based on the use of DFT-optimized geometries as initial guesses for the EXAFS fit and can be easily transposed to the analysis of others metal complexes. Beside the validation of DFT-geometries, the analysis discussed in this section can be regarded as an example case of EXAFS refinement in complex transition metal systems. Here, a careful path selection and grouping is fundamental to correctly model the EXAFS signal, and allows to extract the maximum level of information from EXAFS characterization. Section 5.5.1 discusses the results of the structural refinement in a more discursive approach, while Section 5.5.2 treats the same matter in a more technical and advanced way and can be skipped by beginners.

5.5.1 Structure refinement of *cis*-$[Ru(bpy)_2(py)_2]^{2+}$ in aqueous solution by EXAFS spectroscopy

The DFT optimized geometry of *cis*-$[Ru(bpy)_2(py)_2]^{2+}$ is shown in Figure 5.10a. Due to the complexity of the structure, more than 200 single scattering (SS) and multiple scattering (MS) paths were generated by the FEFF code. Thirty main paths were selected, according to the criteria discussed in detail in Section 5.5.2. These paths were sufficient to properly reconstruct the overall EXAFS signal.

To limit the number of optimized variables, all paths were optimized with the same amplitude factor (S_0^2) and with the same energy shift (ΔE) parameter. Moreover, both the two pyridine (py) and the two bipyridine (bpy) ligands were considered as rigid molecules, whose only degree of freedom is the radial translation along the corresponding Ru–N axis. The two pairs of py and bpy ligands were assumed to behave in the same way, respectively. Consequently, the only two structural parameters optimized in the fit were the distances $R_{Ru-N(py)}$ and $R_{Ru-N(bpy)}$; the lengths of all the other paths were calculated starting from these two values, according to geometrical constraints imposed by the rigidity of the py and bpy units.

Concerning the Debye–Waller (DW) factors, only two parameters were optimized: $\sigma_{Ru-N(py)}$ and $\sigma_{Ru-N(bpy)}$, associated with Ru–N bonds for py or bpy ligands, respectively. DW factors for SS and MS paths involving atoms of the same ligand are modeled as $\sigma^2 = \sigma^2_{Ru-N(L)}(R_{eff}/R_{Ru-N(L)})^{1/2}$ (where L = py or bpy and R_{eff} is the mean free length of the path). In such a way, when R_{eff} increases, the expected proportional expansion of its standard deviation (σ) is roughly accounted. Several almost co-linear MS paths involve two N atoms of two opposite L and L′ ligands: correspondent DW factors were computed as $\sigma^2_{MS} = \sigma^2_{Ru-N(L)} + \sigma^2_{Ru-N(L')}$.

Summarizing, the fit runs over 6 independent parameters only. A similar approach to validate a complex structure by EXAFS (previously refined by XRD of the crystalline phase or optimized *via* DFT calculations), maintaining a limited num-

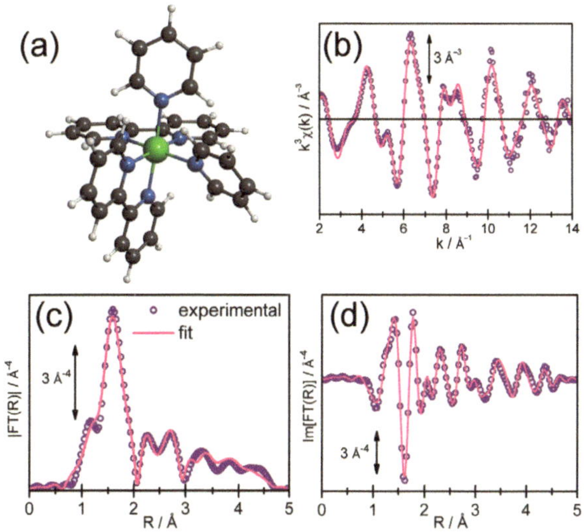

Fig. 5.10: Part (a): DFT optimized geometry of *cis*-$[Ru(bpy)_2(py)_2]^{2+}$. Color code: Ru, green; N, blue, C, gray, H, white. Part (b): experimental $k^3\chi(k)$ EXAFS spectrum of *cis*-$[Ru(bpy)_2(py)_2]^{2+}$ (violet curve) compared with its best fit (magenta curve). Corresponding phase-uncorrected FT modulus in part (b) and part (c) imaginary part. Optimized parameters of the fit are reported in Table 5.3. Unpublished Figure reporting data published in ref. [96].

ber of optimized parameters, has successfully been employed in the similar case of the $[Ru(bpy)_n(AP)_{6-2n}]Cl_2$ (AP = 4-aminopyridine, n = 1, 2, 3) complexes by Salassa *et al.* [96], or of Cp_2Cr hosted in different porous matrices by Estephane *et al.* [37, 83] (and discussed here in Section 5.4), or for metal organic frameworks by Bonino *et al.* [97].

The quality of the obtained fit in both R- and k-spaces can be appreciated in Figure 5.10 and is further confirmed by the very low values of both R-factor and errors associated to the fitting parameters (Tab. 5.3). It is worth highlighting that the bpy units contribute almost twice as much to the high-distance paths as the py units. Consequently, distances and DW factors were determined with a better precision for the bpy ligands. In accordance with their lower mass, py units exhibit a DW factor higher than bpy. The nitrogen atom of the py units is located at a distance of 2.09 ± 0.01 Å from Ru, hence 0.08 Å shorter than the value predicted by DFT calculations (Tab. 5.3). Also the experimental Ru–N(bpy) distances are 0.03 Å shorter than the theoretical ones. Interestingly, the optimized $R_{Ru-N(bpy)}$ and $R_{Ru-N(py)}$ distances become equivalent taking account of the associated errors. Such a trend is characteristic of the EXAFS technique, in particular dealing with structures where several neighbor atoms of the same chemical nature are located at close distances. Here, a tendency towards equalization of the refined bond distances is commonly observed [37, 83, 97–99]. In conclusion, the EXAFS data analysis fully confirms the overall structure optimized in the DFT calculations.

Furthermore, the results reported in Table 5.3 are in good agreement with previous studies performed on similar ruthenium complexes [96, 98, 99].

Tab. 5.3: Summary of the optimized parameters in the EXAFS data fitting (Fig. 5.10). The fits were performed in R-space in the 1.0–5.0 Å range over k^3-weighted FT of the $\chi(k)$ functions performed in the 2.0–14.0 $Å^{-1}$ interval. A single ΔE and a single S_0^2 were optimized for all SS and MS paths. Optimized bond distances are compared to the average values obtained from DFT calculations. Table adapted with permission from [96], copyright American Chemical Society 2013.

EXAFS fit parameters for *cis*-$[Ru(bpy)_2(py)_2]^{2+}$ with comparison to DFT calculations		
Parameter	EXAFS	DFT optimization
N_{ind}	30	–
N_{fit}	6	–
R_{factor}	0.032	–
S_0^2	0.72 ± 0.05	–
ΔE (eV)	2.5 ± 0.6	–
$R_{Ru-N(bpy)}$ (Å)	2.073 ± 0.007	2.11
$\sigma_{Ru-N(bpy)}$ ($Å^2$)	0.0027 ± 0.0006	–
$R_{Ru-N(py)}$ (Å)	2.09 ± 0.01	2.17
$\sigma_{Ru-N(py)}$ ($Å^2$)	0.004 ± 0.001	–

5.5.2 Advanced details of the EXAFS structure refinement of *cis*-$[Ru(bpy)_2(py)_2]^{2+}$

In the EXAFS data, both the $k^3\chi(k)$ function and its Fourier transform in the R-space clearly show a structured signal which reflects the structural complexity of the investigated Ru-complex. As previously mentioned, the DFT-optimized geometry was used as input for the FEFF8.4 code to perform the EXAFS data analysis. Considering that all the atoms of the complex are included in an ideal sphere of radius ~ 5 Å, centered on the Ru atom, the calculation was limited to SS and MS paths with less than four legs and $R_{eff} < 5$ Å. In this way, more than 200 different paths, with their relative amplitudes and phases, were obtained.

Such a high number of paths is due to the low symmetry of the computed structure (geometry optimization performed without symmetry constraint). Indeed, every single path that would be obtained with a perfectly symmetric model is actually split into several slightly different paths (R_{eff} variations in the order of ± 0.01 Å), due to the loss of degeneracy.

Therefore, a significant decrease in the overall number of EXAFS paths can be first obtained by re-grouping together the split paths, involving atoms in quasi-symmetric positions of the molecule. Additionally, a further reduction of the number of considered paths is achieved by excluding those having an amplitude smaller than 10 % of the most intense path (the Ru–N(bpy) SS path, corresponding to the N of the closest bpy unit located at ca. 2.11 Å). Hence, only 30 of the initial 200 SS and MS paths are effectively employed to model the experimental EXAFS spectrum of *cis*-$[Ru(bpy)_2(py)_2]^{2+}$.

Hereafter, the specific contribution of the selected scattering paths to the overall EXAFS signal will be discussed. All paths shown below were computed using the values S_0^2, ΔE, $R_{Ru-N(bpy)}$, $\sigma_{Ru-N(bpy)}$, $R_{Ru-N(py)}$, $\sigma_{Ru-N(py)}$ obtained from the best fit, reported in Table 5.3. Looking at the experimental data (Fig. 5.10b), three well-defined contributions localized in different regions of the R-space can be individuated, in particular in the 0.75–2.05 Å, 2.05–3.00 Å and 3.00–4.75 Å ranges. The first interval contains the first-shell Ru–N SS paths of the four ligands (4- and 2-fold degenerate contributions from bpy and py respectively, light blue curve in Fig. 5.11a,b).

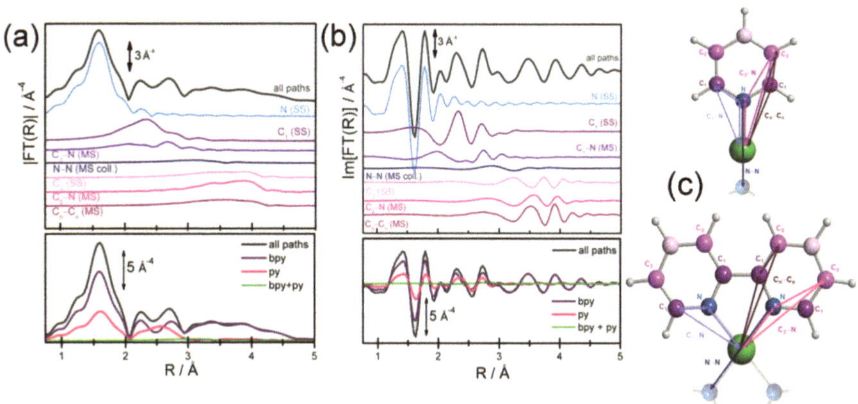

Fig. 5.11: Different path contributions to the overall EXAFS signal of the *cis*-$[Ru(bpy)_2(py)_2]^{2+}$ complex. Part (a) moduli and part (b) imaginary parts of the phase uncorrected FT. All path contributions were computed using the values obtained from the best fit (see Tab. 5.3) for the parameters. The bottom panels show the moduli and imaginary parts (in (a) and (b), respectively) of the phase uncorrected FT relative to the three subsets of paths involving exclusively atoms either of the bpy (purple line) or py ligands (magenta line), or mixed paths involving atoms of both ligand types. Part (c): structure of py and bpy ligands indicating the atoms involved in the EXAFS paths, and the paths themselves using the same color code of part (a). Unpublished Figure reporting data published in ref. [96].

In the 2.05–3.00 Å range, the Ru–C_1 SS and the Ru–C_1–N MS paths contribute (purple and violet lines in Fig. 5.11a, b) with the following degenerations relative to bpy and py: 8 and 4 for the SS, 16 and 8 for the MS. In the MS case, the number of equivalent paths is doubled with respect to SS because the amplitude, phase and total length remain unchanged reversing the order of the atoms involved in the scattering process (Ru–C_1–N $\equiv$ Ru–N–C_1). In the last region, corresponding to higher R-values, the Ru–C_2 SS (light pink curve in Fig. 5.11a,b, degenerated over 12 atoms, 8 belonging to bpy and 4 to py) and other MS paths are present. These involve a C_2 atom with one or more atoms among N (magenta curve), C_1 or a different C_2 (dark pink curves in Fig. 5.11a,b). Furthermore, a Ru–N–N MS collinear path (degeneration 6, dark blue line in Fig. 5.11a,b) is present, which involves three pairs of N atoms in mirror position with respect to the central Ru atom.

Alternatively, the paths can also be grouped in three subsets, depending on the ligand to which each scatterer atom belongs: (i) bpy ligands only, (ii) py ligands only, (iii) both bpy and py ligands. The bottom panels in Figure 5.11 report in R-space (both modulus, panel (a), and imaginary part, panel (b)) the sum of the paths belonging to the three subsets as purple, dark pink, and green curves, respectively. The sum over all the paths, including type (i), (ii), and (iii) signals, is also shown for comparison (gray curve).

From the analysis of the paths' contribution distinguishing by the ligand-type, the following important information emerges:

- the overall contribution of bpy paths (i) is higher than that due to py paths (ii), in agreement with the higher number of atoms contributing to the scattering process;
- the mixed contribution involving atoms of both bpy and py ligands (iii) is limited to one Ru–N–N MS collinear path, 2-fold degenerated, and can be neglected;
- in the higher R-values interval (3.00–4.75 Å), the ratio between signal amplitude of (i) and (ii) shows a remarkable change. The 2:1 ratio present in the smaller R-intervals now becomes notably higher. This could be explained by considering that for the bpy units more MS paths are possible, since the atoms of two different rings linked together are involved. Furthermore, the bidentate nature of the bpy ligand is such that the Ru–N–C_2 angles are closer to 180° with respect to the same angles in the py ligand (Fig. 5.11c), and consequently the corresponding paths are closer to collinearity.

5.6 EXAFS study on MOFs of the UiO-66/UiO-67 family: comparison with XRPD and ab initio investigations

The recently discovered UiO-66/67/68 class of isostructural MOFs [100] has attracted great interest because of its remarkable stability at high temperatures, high pressures and in presence of different solvents acids and bases [101].

UiO-66 is obtained by connecting $Zr_6O_4(OH)_4$ inorganic cornerstones with 1,4-benzene-dicarboxylate (BDC) as linker, while the isostructural UiO-67 material is obtained using the longer 4,4'-biphenyl-dicarboxylate (BPDC) linker [102] (Fig. 5.12a) and Hf-UiO-66 is obtained by keeping the UiO-66 linker (BDC) and substituting the $Zr_6O_4(OH)_4$ blocks with $Hf_6O_4(OH)_4$ corners (inset in Fig. 5.12e). XRPD (Fig. 5.12b,e) testifies the quality of the synthesis. Due to the rigidity of the framework, several isostructural UiOs have been prepared and tested for stability and gas adsorption. Kandiah *et al.* [103] studied the thermal and chemical stabilities of isostructural UiO-66-X (X = NH_2, Br and NO_2) and observed the lower stability of this analogue with respect to parent UiO-66. Conversely, as documented by the thermogravime-

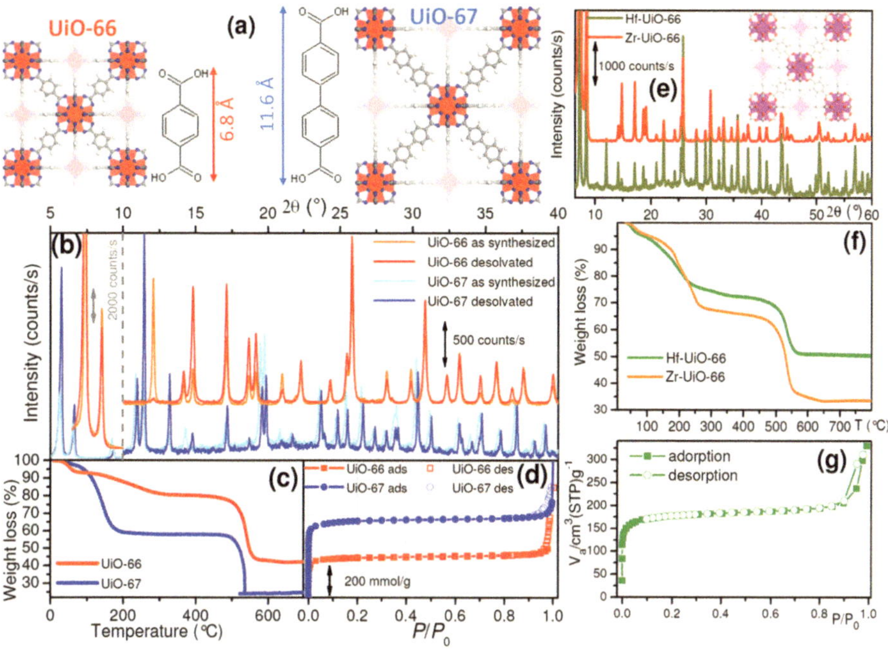

Fig. 5.12: Part (a): From left to right: comparison of the dimensions of linker and structure for the isostructural UiO-66 and UiO-67 MOFs. Part (b): comparison of the XRPD patterns (λ = 1.540 Å) for prepared UiO-66 (UiO-67), orange (cyan) curve and activated at 300° C (red and blue curves). The patterns in the 10–40 2θ° region have been amplified by a factor of 4. Patterns related to UiO-66 have been vertically translated for clarity. Part (c): TGA curve of UiO-67 and UiO-67 samples, red and blue curves, respectively. In both cases, the heating ramp was of 5 C/min in a N_2 flow (100 ml/min). Part (d): volumetric N_2 adsorption isotherms recorded at 77 K on UiO-66 (red squares) and UiO-67 (blue circles). Empty and filled scatters refer to the adsorption and desorption branches, respectively. Part (e): XRPD pattern (λ = 1.540 Å) of Hf-UiO-66 (green) and Zr-UiO-66 (red) in their solvated forms. The inset reports the MOF structure. Part (f): weight loss of Hf-UiO-66 relative to the start mass (green curve). The weight loss of Zr-UiO-66 has been added as a reference (orange curve). Since hafnium is 41 % heavier than zirconium, the Zr-UiO-66 shows both higher initial and breakdown losses. Part (g): N_2 adsorption/desorption isotherm for Hf-UiO-66 at 77 K. Unpublished figure reporting data from [100–102, 104, 105].

try studies reported in Figure 5.12c,f, UiO-67 [102] and Hf-UiO-66 [104] show thermal and chemical stability similar to that of UiO-66 and exhibit the expected surface area, as determined by low temperature volumetric N_2 adsorption isotherms (Fig. 5.12d,g). Such a high stability is related to the fact that each Zr- (Hf-) octahedron is 12-fold connected to adjacent octahedra. This connectivity is very common for metals, resulting in the highly packed fcc structure, but it is still almost unique in MOF topologies.

The desolvation process left the XRPD pattern of such materials almost unchanged (Fig. 5.12b): besides a gain in intensity of the basal reflections (due to the removal of

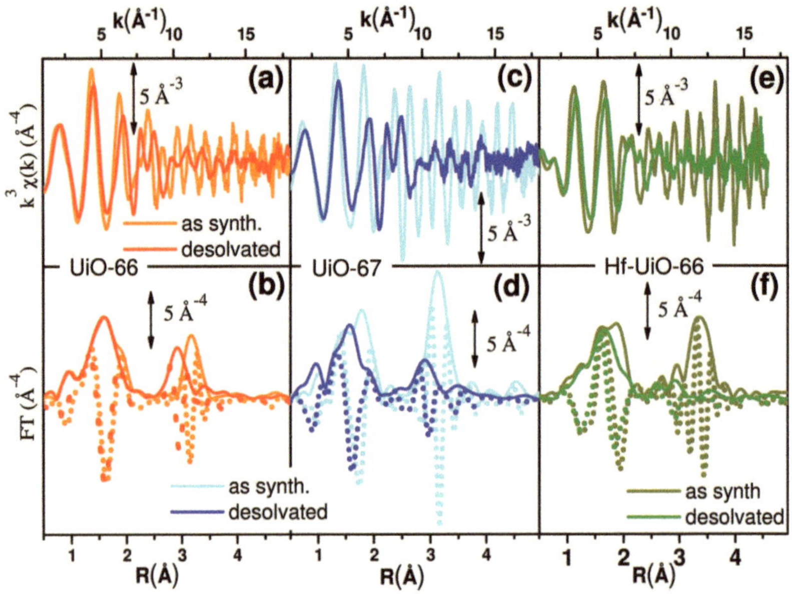

Fig. 5.13: *k*-space (top panels) and R-space (bottom panels) EXAFS data collected on UiO-66, UiO-67 and Hf-UiO-66, parts (a, b), (c, d) and (e, f), respectively. Both synthesized (or hydroxylated) and desolvated (or dehydroxylated) forms of the three different isostructural MOFs have been measured. With the exception of the desolvated Hf-UiO-66 sample (collected at 573 K) remaining spectra were collected at 300 K. Unpublished figure reporting data from [100–102, 104, 105].

electron density inside the pores) [106–108] all peaks remain in almost the same 2θ position with small intensity changes. Conversely, a huge modification of the EXAFS spectra is obtained in all cases, see Figure 5.13.

In the three hydroxylated materials, the structure determined from the Rietveld refinement of the XRPD corresponding patterns resulted in a straightforward interpretation of the complex EXAFS signals, see the first three columns in Table 5.4. The dramatic modification undergone by the EXAFS spectrum upon dehydroxylation (Fig. 5.13) makes the data analysis not so straightforward. In the case of UiO-66 (see Fig. 5.13b, but similar effects are observed in the two other cases) the changes are basically explained in terms of three main effects: (i) small contraction of the first M-O shell (M = Zr or Hf) accompanied by a small decrease in coordination (erosion of the shoulder around 1.9 Å); (ii) relevant distortion of the second shell contribution showing a maximum that moves from 3.17 Å to 2.91 Å, with a shoulder at 3.41 Å, thus reflecting an important splitting of the R_{M1} distances of the octahedron sides; (iii) the almost complete disappearance of the weak contribution around 4.7 Å, due to the M-M SS signal of the octahedron diagonal (R_{M2}). For the three cases, differently to the hydroxylated cases, the 3D model obtained from the Rietveld refinement of XRPD data in the highly symmetric *Fm-3m* space group was inadequate to simulate the experimental datum. The origin of this failure was obviously due to the inability of the model to account for two different R_{M1} and R_{M2} distances. For both the UiO-66 [101] and the UiO-67 [102] cases, the failure of the XRPD model was overcome by using the optimized geometry obtained by *ab initio* periodic calculations.

In presence of the solvent in the pores, the inorganic cornerstones are perfect $M_6(OH)_4O_4$ octahedron (see model in Fig. 5.14b), with 6 equivalent M at the vertex, 12 equivalent M-M1 sides and 3 equivalent and M-M2 diagonals. Upon desolvation, 2 structural water molecules are lost per cornerstone unit (Fig. 5.14a) that evolves from $M_6(OH)_4O_4$ to M_6O_6 [101, 102, 104, 105]. The new M_6O_6 octahedron is compressed (2 opposite vertexes approaching, see model in Fig. 5.14c) resulting in the shortening of 8 of the 12 edges, and the elongation of the other 4 edges. To take into account this variation we simulated the EXAFS contribution with two independently parameterized paths fixing for the degeneration a ratio of 1/3 and 2/3 with respect to the case of the single contribution. For the three systems, this combined XRPD, EXAFS and DFT approach allowed a full interpretation of the EXAFS data in both hydroxylated and dehydroxylated forms [101, 102, 104, 105]. Please note that IR spectroscopy was crucial to confirm this model as it allowed to observe the disappearance of the O-H stretching band in these materials.

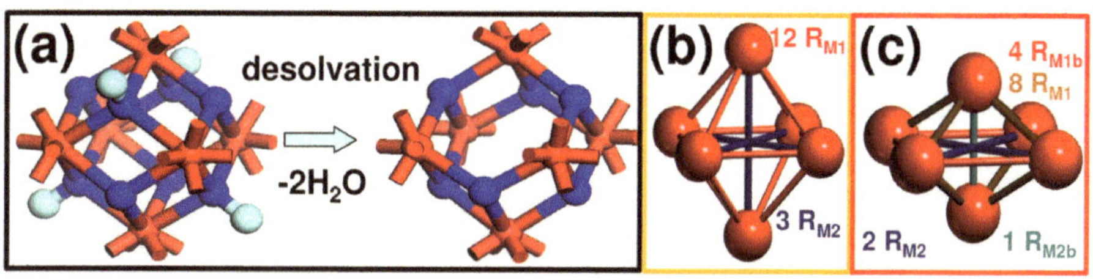

Fig. 5.14: Part (a): Stick and ball representation of the dehydroxylation undergone by the inorganic $M_6O_4(OH)_4$ cornerstone upon thermal treatment at 300 °C in vacuo resulting in a distorted M_6O_6 cluster (M = Zr or Hf). Red, blue and cyan colors refer to M, O, and H atoms, respectively. Part (b): stick and ball representation of the perfect M_6 octahedron, showing 12 equivalent R_{M1} sides and 3 equivalent $R_{M2} = \sqrt{2}R_{M1}$ diagonals. Part (c): stick and ball representation of a squeezed M_6 octahedron. The 12 sides are now split into 4 in-plane long M_{M1b} sides and 8 prismatic short M_{M1a} sides, while the 3 diagonals evolve into 2 in-plane long M_{M2b} and 1 orthogonal short M_{M2a} diagonals. For clarity, O atoms are omitted in parts (b) and (c). Unpublished figure.

Summarizing, EXAFS spectroscopy allowed to detect the evolution from $M_6(OH)_4O_4$ to M_6O_6 (M = Zr or Hf) of the inorganic cornerstones of UiO-66, UiO-67 and Hf-UiO-66 MOFs occurring in the desolvation process that had escaped XRPD detection. On Zr-UiO-66 and Zr-UiO-67, periodic calculations performed with CRYSTAL code [109] at DFT level of theory support EXAFS data.

Tab. 5.4: Summary of the EXAFS refinement obtained on the hydroxylated and dehydroxylated forms of UiO-66, UiO-67 and Hf-UiO-66. Parameters without error bars were not optimized. The EXAFS refinement of the hydroxylated materials was obtained using as input model the optimized structure from Rietveld refinement of the corresponding XRPD patterns. The EXAFS refinement of the dehydroxylated materials was obtained using as input model optimized *ab initio* calculations for the hydroxylated of UiO-66. With this approach the coordination number (N) of each contribution is fixed by the model stoichiometry. Refinement of the experimental amplitude is done by optimizing the overall amplitude factor S_0^2 only. The fitting of the higher shells was possible only adopting the axial compressed model of the M6O6 octahedron represented where eight octahedron sides R_{M1} are split into eight short prismatic distances (R_{M1}, N = 8/3) and four long planar ones (R_{M1b}, N = 4/3) and where the three diagonals R_{M2} are split into a short axial diagonal and (R_{M2a}, involving two M atoms out of six; N = 1/3) and two long planar diagonals (R_{M2b}, involving four M atoms out of six; N = 2/3): Fig. 5.14c. Unpublished table, reporting data from [100, 101, 102, 104, 105].

	UiO-66	Hydro UiO-67 hydroxylated	Hf-UiO-66	UiO-66	UiO-67 dehydroxylated	Hf-UiO-66
T (K)	300	300	300	300	300	573
R-factor	0.01	0.04	0.02	0.02	0.04	0.05
Δk ($Å^{-1}$)	2.0-18.0	2.0-16.2	2.0-16.0	2.0-15.0	2.0-15.0	2.0-15.0
ΔR (Å)	1.0-5.3	1.0-5.3	1.0-5.5	1.0-5.3	1.0-5.3	1.0-3.9
Indip. points	43	38	40	35	35	25
N. variables	14	13	10	15	15	9
ΔE_0 (eV)	5 ± 1	1 ± 1	2.2 ± 0.6	5	1	2.4 ± 0.7
S_0^2	1.17 ± 0.08	1.17	0.91 ± 0.06	1.17	1.17	0.91
$R_{\mu 3-O}$ (Å)	2.087 ± 0.008	2.12 ± 0.02	2.12 ± 0.01	2.06 ± 0.01	2.096 ± 0.007	2.06 ± 0.01
$\sigma^2(\mu_3\text{-O})$ ($Å^2$)	0.0036 ± 0.0009	0.005 ± 0.002	0.005 ± 0.002	0.008 ± 0.003	0.006 ± 0.001	0.009 ± 0.002
R_{O1} (Å)	2.235 ± 0.008	2.26 ± 0.01	2.25 ± 0.01	2.221 ± 0.007	2.249 ± 0.007	2.19 ± 0.01
σ^2(O1) ($Å^2$)	0.0074 ± 0.0008	0.006 ± 0.001	0.005	0.007 ± 0.002	0.004 ± 0.001	0.009
R_C (Å)	3.19 ± 0.02	3.40 ± 0.06	3.23 ± 0.06	3.17 ± 0.04	$3.15 \pm 0.04)$	3.22 ± 0.05
σ^2(C) ($Å^2$)	0.004 ± 0.002	0.012 ± 0.002	0.014 ± 0.011	0.009 ± 0.009	0.004 ± 0.003	0.016 ± 0.013
R_{M1} (Å)	3.511 ± 0.007	3.512 ± 0.006	3.510 ± 0.005	3.35 ± 0.01	3.365 ± 0.015	3.31 ± 0.03
σ^2(M1) ($Å^2$)	0.007 ± 0.001	0.004 ± 0.001	0.0042 ± 0.0004	0.009 ± 0.001	0.009 ± 0.002	0.009 ± 0.004
R_{M1b} (Å)	–	–	–	3.74 ± 0.02	3.80 ± 0.03	3.45 ± 0.06
σ^2(M1b) ($Å^2$)	–	–	–	0.009 ± 0.002	0.008 ± 0.003	0.009
R_{M2} (Å)	4.99 ± 0.04	4.95 ± 0.03	4.964	$4.14 \pm 0.07)$	4.15 ± 0.07	–
σ^2(M2) ($Å^2$)	0.010 ± 0.006	0.004 ± 0.002	0.008 ± 0.002	0.008 ± 0.006	0.006 ± 0.004	–
R_{M2b} (Å)	–	–	–	5.30 ± 0.04	5.46 ± 0.05	–
σ^2(M2b) ($Å^2$)	–	–	–	0.008	0.006	–

5.7 Applications of X-ray micro beams: Electroabsorption modulated laser for optoelectronic devices

Synchrotron radiation X-ray micro- and nano-beams are emerging characterization tools with broad implications for science, ranging from solid state physics to structural biology. In recent years space-resolved characterization at the (sub-)micrometric scale was made possible by the exponential increase in X-ray brilliance at III generation synchrotron radiation sources and by the impressive improvements in X-ray focusing devices [110]. In the field of materials characterization, they are becoming a key tool for the space-resolved determination of structural (XRD) [111–114] and electronic (XANES/EXAFS) properties [115, 116] and for chemical speciation (XRF) [117–120] of nanostructured or composite materials.

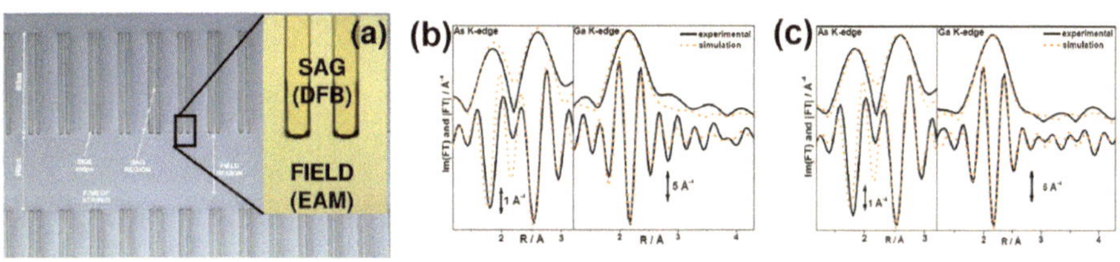

Fig. 5.15: Part (a): Optical micrograph of the InP substrate patterned with SiO_2 stripes which allow SAG growth. Part (b): k^3-weighed, phase uncorrected, Fourier transform modulus and imaginary part of Ga and As K-edges, performed in the SAG region. For the fit a co-refinement approach was adopted. Part (c): as part (b) for the FIELD region. Unpublished figure reporting data published in [116].

In particular, synchrotron radiation micro-/nano-beams can also be successfully employed to study devices based on nanostructured inorganic materials. An interesting example is represented by the electroabsorption modulated laser (EML) [121, 122]. Indeed, nowadays heterostructures based on quaternary III-V semiconductor alloys are widely used in optical communication systems. Advanced optoelectronic devices often require the integration of two different functions in the same chip: excellent results in the development of monolithic integration have been reached with the selective area growth (SAG) technique [123, 124]. SAG exploits the perturbation of the growth fluxes induced by a dielectric mask: when the metallorganic precursors collide with the dielectric mask, they are deflected and they can migrate through the unmasked semiconductor where the growth can start. In this way the reactive species coming from the gas phase are enriched by those deflected by the mask: the result is a variation in composition and thickness of semiconductors grown near (SAG region) and far (FIELD region) from the mask (Fig. 5.15a). The EML, obtained by monolithic integration of an electroabsorption modulator (EAM) with a distributed feedback laser (DFB), is one of the most promising applications of SAG. A voltage modulation applied to the

EAM switches it between an opaque and a transparent state by means of the Stark effect and ensures the modulation of the DFB laser emission, allowing long-distance communications (up to 80 km) at high frequency (10 Gb/s).

Mino *et al.* [116] investigated a SAG EML device based on $Al_{xw}Ga_{yw}In_{1\text{-}xw\text{-}yw}As$/ $Al_{xb}Ga_{yb}In_{1\text{-}xb\text{-}yb}As$ (compressive-strained well / tensile-strained barrier) MQW structures grown on InP by metallorganic vapor phase epitaxy. They reported μ-EXAFS spectra acquired at the ESRF ID22 beamline employing a pair of Kirkpatrick–Baez (KB) mirrors to reach a beam size of 1.7 μm (vertical) × 5.3 μm (horizontal), which allowed them to study in a space-resolved way the atomic local environment in the SAG and FIELD regions.

Tab. 5.5: Summary of the parameters optimized in the fitting of the EXAFS data (Fig. 5.15b,c) on the SAG and FIELD regions (Fig. 5.15a). A corefinement approach on As and Ga K-edges was adopted. Unpublished table reporting data published in [116].

	SAG	FIELD
σ^2_{As-Ga} (Å^2)	0.006 ± 0.002	0.005 ± 0.002
σ^2_{As-In} (Å^2)	0.008 ± 0.003	0.007 ± 0.003
σ^2_{As-Al} (Å^2)	0.010 ± 0.004	0.008 ± 0.004
$R_{As\text{-}In}$ (Å)	2.60 ± 0.02	2.60 ± 0.02
ΔR_{As-In} (Å)	−0.02 ± 0.02	−0.02 ± 0.02
R_{As-Al} (Å)	2.49 ± 0.11	2.48 ± 0.11
$R_{As\text{-}Ga}$ (Å)	2.469 ± 0.007	2.463 ± 0.005
ΔR_{As-Ga} (Å)	0.021 ± 0.007	0.015 ± 0.005

By performing a combined fit of Ga and As K-edges spectra (Fig. 5.15b,c and Tab. 5.5), they observed a stretching of the As-Ga first shell bond distances and a contraction of the As-In first shell distances with respect to the binary compounds InAs and GaAs. These distances are in agreement with experimental [125] and theoretical [126] data on strained $In_{0.60}Ga_{0.40}$ As alloys on InP. Moreover, they noticed that $\Delta R_{As\text{-}Ga}$ is higher in the SAG than in the FIELD region and a higher structural disorder (monitored by the σ^2 values) is observed in the SAG region, however the quality of the data resulted in too high error bars that did not allow them to reliably discriminate between SAG and FIELD regions (Tab. 5.5).

Note added in proofs

In Section 5.5 we reported the EXAFS study of the ground state of the *cis*-$[Ru(bpy)_2(py)_2]^{2+}$ complex. Recently, Borfecchia *et al.* [127] have been able to highlight the structure of the ^{3}MLCT excited-state structure of the complex with a 100-ps resolution time-resolved EXAFS experiment (laser pump–X-ray probe) performed at the

11-ID-D beamline of the APS. Data have been analyzed with the innovative approach of the differential EXAFS analysis [128]. This study complement the previous X-ray scattering study performed by the same group [96].

Bibliography

[1] Margaritondo G. Introduction to Synchrotron Radiation. New York, Oxford, 1988.

[2] Wille K. Synchrotron radiation sources. Rep. Prog. Phys. 1991; 54: 1005–1068.

[3] Bilderback DH, Elleaume P, Weckert E. Review of third and next generation synchrotron light sources. J. Phys. B-At. Mol. Opt. Phys. 2005; 38: S773–S797.

[4] Couprie ME, Filhol JM. X radiation sources based on accelerators. C. R. Phys. 2008; 9: 487–506.

[5] Sandstrom DR, Lytle FW. Developments in extended X-ray absorption fine structure applied to chemical systems. Ann. Rev. Phys. Chem. 1979; 30: 215–238.

[6] Lee PA, Citrin PH, Eisenberger P, Kincaid M. Extended X-ray absorption fine structure – its strengths and limitations as structural tool. Rev. Mod. Phys. 1981; 53: 769–806.

[7] Teo BK, Joy DC. EXAFS Spectroscopy: Techniques and Applications New York, Plenum, 1981.

[8] Stern EA. Theory of EXAFS, in: Koningsberger DC, Prins R (eds.) X-Ray Absorption: Principles, Applications, Techniques of EXAFS, SEXAFS and XANES. pp. 3–51, New York: John Wiley & Sons, 1988.

[9] Filipponi A, Di Cicco A, Natoli CR. X-ray-absorption spectroscopy and n-body distribution functions in condensed matter 1. Theory. Phys. Rev. B 1995; 52: 15122–15134.

[10] Filipponi A, Di Cicco A. X-ray-absorption spectroscopy and n-body distribution functions in condensed matter 2. Data analysis and applications. Phys. Rev. B 1995; 52: 15135–15149.

[11] Rehr JJ, Albers RC. Theoretical approaches to x-ray absorption fine structure. Rev. Mod. Phys. 2000; 72: 621–654.

[12] Boscherini F. X-ray absorption fine structure in the study of semiconductor heterostructures and nanostructures, in: Lamberti C (ed.) Characterization of Semiconductor Heterostructures and Nanostructures, pp. 289–330, Amsterdam: Elsevier, 2008.

[13] Bordiga S, Groppo E, Agostini G, van Bokhoven JA, Lamberti C. Reactivity of surface species in heterogeneous catalysts probed by *in situ* X-ray absorption techniques. Chem. Rev. 2013; 113: 1736–1850.

[14] Bart JCJ, Vlaic G. Extended X-ray absorption fine-structure studies in catalysis. Adv. Catal. 1987; 35: 1–138.

[15] Sinfelt JH, Meitzner GD. X-ray absorption-edge studies of the electronic-structure of metal-catalysts. Accounts Chem. Res. 1993; 26: 1–6.

[16] Newton MA, Dent AJ, Evans J. Bringing time resolution to EXAFS: recent developments and application to chemical systems. Chem. Soc. Rev. 2002; 31: 83–95.

[17] Bare SR, Ressler T. Characterization of catalysts in reactive atmospheres by X-ray absorption spectroscopy. Adv. Catal. 2009; 52: 339–465.

[18] Alexeev O, Gates BC. EXAFS characterization of supported metal-complex and metal-cluster catalysts made from organometallic precursors. Top. Catal. 2000; 10: 273–293.

[19] Sinfelt JH. Structure of bimetallic clusters. Accounts Chem. Res. 1987; 20: 134–139.

[20] Frenkel A. Applications of extended X-ray absorption fine-structure spectroscopy to studies of bimetallic nanoparticle catalysts. Chem. Soc. Rev. 2012; 41: 7965–7965.

[21] Abruna HD. Probing electrochemical interfaces with X-rays. Adv. Chem. Phys. 1990; 77: 255–335.

[22] Abruna HD, Bommarito GM, Yee HS. X-ray standing waves and surface EXAFS studies of electrochemical interfaces. Accounts Chem. Res. 1995; 28: 273–279.

[23] Bordiga S, Bonino F, Lillerud KP, Lamberti C. X-ray absorption spectroscopies: useful tools to understand metallorganic frameworks structure and reactivity. Chem. Soc. Rev. 2010; 39: 4885–4927.

[24] Riggsgelasco PJ, Stemmler TL, Pennerhahn JE. XAFS of dinuclear metal sites in proteins and model compounds. Coord. Chem. Rev. 1995; 144: 245–286.

[25] Pellerito L, Nagy L. Organotin(IV)$^{n+}$ complexes formed with biologically active ligands: equilibrium and structural studies, and some biological aspects. Coord. Chem. Rev. 2002; 224: 111–150.

[26] Penner-Hahn JE. Characterization of "spectroscopically quiet" metals in biology. Coord. Chem. Rev. 2005; 249: 161–177.

[27] Strange RW, Ellis M, Hasnain SS. Atomic resolution crystallography and XAFS. Coord. Chem. Rev. 2005; 249: 197–208.

[28] Denecke MA. Actinide speciation using X-ray absorption fine structure spectroscopy. Coord. Chem. Rev. 2006; 250: 730–754.

[29] Chen JG. NEXAFS investigations of transition metal oxides, nitrides, carbides, sulfides and other interstitial compounds. Surf. Sci. Rep. 1997; 30: 5–152.

[30] Lamberti C. The use of synchrotron radiation techniques in the characterization of strained semiconductor heterostructures and thin films. Surf. Sci. Rep. 2004; 53: 1–197.

[31] Le Toquin R, Paulus W, Cousson A, Prestipino C, Lamberti C. Time-resolved in situ studies of oxygen intercalation into $SrCoO_{2.5}$, performed by neutron diffraction and X-ray absorption spectroscopy. J. Am. Chem. Soc. 2006; 128: 13161–13174.

[32] Fermi E. Eine statistische Methode zur Bestimmung einiger Eigenschaften des Atoms und ihre Anwendung auf die Theorie des periodischen Systems der Elemente. Z. Phys. 1928; 48: 73–79.

[33] Fermi E. Nuclear Physics. Chicago, University of Chicago Press, 1950.

[34] Borfecchia E, Garino C, Salassa L, Lamberti C. Synchrotron ultrafast techniques for photoactive transition metal complexes. Phil. Trans. R. Soc. A 2013: in press, doi:10.1098/rsta.2012.0132.

[35] Sayers DE, Stern EA, Lytle FW. New technique for investigating non-crystalline structures: Fourier analysis of the extended X-ray absorption fine structure. Phys. Rev. Lett. 1971; 27: 1204–1207.

[36] Lytle FW, Sayers DE, Stern EA. Extended X-ray-absorption fine-structure technique. II. Experimental practice and selected results. Phys. Rev. B 1975; 11: 4825–4835.

[37] Estephane J, Groppo E, Damin A, Vitillo JG, Gianolio D, Lamberti C, Bordiga S, Prestipino C, Nikitenko S, Quadrelli EA, Taoufik M, Basset JM, Zecchina A. Structure and enhanced reactivity of chromocene carbonyl confined inside cavities of NaY zeolite. J. Phys. Chem. C 2009; 113: 7305–7315.

[38] Silversmit G, Vekemans B, Nikitenko S, Bras W, Czhech V, Zaray G, Szaloki I, Vincze L. Polycapillary-optics-based micro-XANES and micro-EXAFS at a third-generation bending-magnet beamline. J. Synchrot. Radiat. 2009; 16: 237–246.

[39] Westre TE, Kennepohl P, DeWitt JG, Hedman B, Hodgson KO, Solomon EI. A multiplet analysis of Fe K-edge 1s $\rightarrow$ 3d pre-edge features of iron complexes. J. Am. Chem. Soc. 1997; 119: 6297–6314.

[40] Koningsberger DC, Prins R. X-ray Absorption: Principles, Applications, Techniques of EXAFS, SEXAFS, and XANES. New York, John Wiley & Sons, 1988.

[41] Harris DC, Bertolucci MD. Symmetry and Spectroscopy. An Introduction to Vibrational and Electronic Spectroscopy. New York, Dover Publications, 1989.
[42] Lamberti C, Prestipino C, Bonino F, Capello L, Bordiga S, Spoto G, Zecchina A, Moreno SD, Cremaschi B, Garilli M, Marsella A, Carmello D, Vidotto S, Leofanti G. The chemistry of the oxychlorination catalyst: An in situ, time-resolved XANES study. Angew. Chem. Int. Edit. 2002; 41: 2341–2344.
[43] Muddada NB, Olsbye U, Caccialupi L, Cavani F, Leofanti G, Gianolio D, Bordiga S, Lamberti C. Influence of additives in defining the active phase of the ethylene oxychlorination catalyst. Phys. Chem. Chem. Phys. 2010; 12: 5605–5618.
[44] Muddada NB, Olsbye U, Leofanti G, Gianolio D, Bonino F, Bordiga S, Fuglerud T, Vidotto S, Marsella A, Lamberti C. Quantification of copper phases, their reducibility and dispersion in doped-$CuCl_2/Al_2O_3$ catalysts for ethylene oxychlorination. Dalton Trans. 2010; 39: 8437–8449.
[45] Piovano A, Agostini G, Frenkel AI, Bertier T, Prestipino C, Ceretti M, Paulus W, Lamberti C. Time resolved in situ XAFS study of the electrochemical oxygen intercalation in $SrFeO_{2.5}$ Brownmillerite structure: Comparison with the homologous $SrCoO_{2.5}$ system. J. Phys. Chem. C 2011; 115: 1311–1322.
[46] Lamberti C, Palomino GT, Bordiga S, Berlier G, D'Acapito F, Zecchina A. Structure of homoleptic $Cu^I(CO)_3$ cations in Cu^I exchanged ZSM-5 zeolite: An X-ray absorption study. Angew. Chem. Int. Edit. 2000; 39: 2138–2141.
[47] Ricchiardi G, Damin A, Bordiga S, Lamberti C, Spanò G, Rivetti F, Zecchina A. Vibrational structure of Titanium silicate catalysts. A spectroscopic and theoretical study. J. Am. Chem. Soc. 2001; 123: 11409–11419.
[48] Belli M, Scafati A, Bianconi A, Mobilio S, Palladino L, Reale A, Burattini E. X-ray absorption near edge structures (XANES) in simple and complex Mn compounds. Solid State Commun. 1980; 35: 355–361.
[49] Bianconi A, Bauer RS. Evidence of SiO at the Si-oxide interface by surface soft X-ray absorption near edge spectroscopy. Surf. Sci. 1980; 99: 76–86.
[50] Oudenhuijzen MK, van Bokhoven JA, Miller JT, Ramaker DE, Koningsberger DC. Three-site model for hydrogen adsorption on supported platinum particles: Influence of support ionicity and particle size on the hydrogen coverage. J. Am. Chem. Soc. 2005; 127: 1530–1540.
[51] Bianconi A, Doniach S, Lublin D. X-ray Ca K edge of calcium adenosine triphosphate system and of simple Ca compounds. Chem. Phys. Lett. 1978; 59: 121–124.
[52] Bressler C, Chergui M. Ultrafast X-ray absorption spectroscopy. Chem. Rev. 2004; 104: 1781–1812.
[53] Stern EA. Theory of the extended x-ray-absorption fine structure. Phys. Rev. B 1974; 10: 3027–3037.
[54] Teo B-K, Lee PA. Ab initio calculations of amplitude and phase functions for extended x-ray absorption fine structure spectroscopy. J. Am. Chem. Soc. 1979; 101: 2815–2832.
[55] McKale AG, Veal BW, Paulikas AP, Chan SK, Knapp GS. Improved ab initio calculations of amplitude and phase functions for extended x-ray absorption fine structure spectroscopy. J. Am. Chem. Soc. 1988; 110: 3763–3768.
[56] For a compilation of electron inelastic mean free path lengths (λ) in solids for energies in the range 0–10 000 eV above the Fermi level, see e.g. Seah MP, Dench WA. Quantitative electron spectroscopy of surfaces: A standard data base for electron inelastic mean free paths in solids. Surf. Interface Anal., 1979; 1: 2–11.
[57] Filipponi A. EXAFS for liquids. J. Phys.-Condes. Matter 2001; 13: R23–R60.
[58] Debye P. Interferenz von Röntgenstrahlen und Wärmebewegung. Ann. Phys. 1913; 348: 49–92.

[59] Waller I. Zur Frage der Einwirkung der Wärmebewegung auf die Interferenz von Röntgenstrahlen. Z. Phys. 1923; 17: 398–408.
[60] Asakura K. Analysis of EXAFS, in: Iwasawa Y (ed.), X-ray absorption fine structure for catalysts and surfaces, pp. 33–58, Singapore: World Scientific, 1996.
[61] Nyquist H. Certain topics in telegraph transmission theory. Trans. AIEE 1928; 47: 617–644. Reprinted in Proc. IEEE 2002; 90: 280–305.
[62] Shannon CE. Communication in the presence of noise. Proc. Institute Radio Engin. 1949; 37: 10–21. Reprinted in Proc. IEEE 1988; 1986: 1447.
[63] Newmann MN. Encyclopedia of Polymer Science and Engineering. New York, Wiley, 1985.
[64] Leofanti G, Marsella A, Cremaschi B, Garilli M, Zecchina A, Spoto G, Bordiga S, Fisicaro P, Berlier G, Prestipino C, Casali G, Lamberti C. Alumina-supported copper chloride 3. Effect of exposure to ethylene. J. Catal. 2001; 202: 279–295.
[65] Leofanti G, Marsella A, Cremaschi B, Garilli M, Zecchina A, Spoto G, Bordiga S, Fisicaro P, Prestipino C, Villain F, Lamberti C. Alumina-supported copper chloride 4. Effect of exposure to O_2 and HCl. J. Catal. 2002; 205: 375–381.
[66] Leofanti G, Padovan M, Garilli M, Carmello D, Marra GL, Zecchina A, Spoto G, Bordiga S, Lamberti C. Alumina-supported copper chloride 2. Effect of aging and thermal treatments. J. Catal. 2000; 189: 105–116.
[67] Leofanti G, Padovan M, Garilli M, Carmello D, Zecchina A, Spoto G, Bordiga S, Palomino GT, Lamberti C. Alumina-supported copper chloride 1. Characterization of freshly prepared catalyst. J. Catal. 2000; 189: 91–104.
[68] Prestipino C, Bordiga S, Lamberti C, Vidotto S, Garilli M, Cremaschi B, Marsella A, Leofanti G, Fisicaro P, Spoto G, Zecchina A. Structural determination of copper species on the alumina-supported copper chloride catalyst: A detailed EXAFS study. J. Phys. Chem. B 2003; 107: 5022–5030.
[69] Joyner RW, Martin KJ, Meehan P. Some applications of statistical tests in analysis of EXAFS and SEXAFS Data. J. Phys. C: Solid State Phys. 1987; 20: 4005–4012.
[70] Lee PA, Beni G. New method for the calculation of atomic phase shifts: Application to extended x-ray absorption fine structure (EXAFS) in molecules and crystals. Phys. Rev. B 1977; 15: 2862–2883.
[71] Muddada NB, Olsbye U, Fuglerud T, Vidotto S, Marsella A, Bordiga S, Gianolio D, Leofanti G, Lamberti C. The role of chlorine and additives on the density and strength of Lewis and Brønsted acidic sites of γ-Al_2O_3 support used in oxychlorination catalysis: A FTIR study. J. Catal. 2012; 284: 236–246.
[72] Gianolio D, Muddada NB, Olsbye U, Lamberti C. Doped-$CuCl_2/Al_2O_3$ catalysts for ethylene oxychlorination: Influence of additives on the nature of active phase and reducibility. Nucl. Instrum. Met. Phys. Res. B 2012; 284: 53–57.
[73] Lamberti C, Bordiga S, Bonino F, Prestipino C, Berlier G, Capello L, D'Acapito F, Xamena FXLI, Zecchina A. Determination of the oxidation and coordination state of copper on different Cu-based catalysts by XANES spectroscopy in situ or in operando conditions. Phys. Chem. Chem. Phys. 2003; 5: 4502–4509.
[74] Coperet C, Chabanas M, Saint-Arroman RP, Basset JM. Homogeneous and heterogeneous catalysis: Bridging the gap through surface organometallic chemistry. Angew. Chem. Int. Edit. 2003; 42: 156–181.
[75] Davis ME. Ordered porous materials for emerging applications. Nature 2002; 417: 813–821.
[76] Ozin GA, Gil C. Intrazeolite organometallics and coordination-complexes – internal versus external confinement of metal guests. Chem. Rev. 1989; 89: 1749–1764.

[77] Karol FJ, Brown GL, Davison JM. Chromocene-based catalysts for ethylene polymerization: Kinetic parameters. J. Polym. Sci. A 1973; 11: 413–424.
[78] Thomas JL, Brintzinger HH. Molybdenocene as an intermediate in reactions with hydrogen, nitrogen, and carbon monoxide. J. Am. Chem. Soc. 1972; 94: 1386–1387.
[79] Tang Wong KL, Brintzinger HH. Reactivity patterns of chromocene, molybdenocene, and tungstenocene reaction systems. I. Carbonyl complex formation as a probe of coordinative unsaturation. J. Am. Chem. Soc. 1975; 97: 5143–5146.
[80] Brintzinger HH, Lohr LL, Tang Wong KL. Reactivity patterns of chromocene, molybdenocene, and tungstenocene reaction systems. II. Analysis in terms of molecular orbital and electron-electron repulsion energies. J. Am. Chem. Soc. 1975; 97: 5146–5155.
[81] Estephane J, Groppo E, Vitillo JG, Damin A, Lamberti C, Bordiga S, Zecchina A. Chromocene in porous polystyrene: An example of organometallic chemistry in confined spaces. Phys. Chem. Chem. Phys. 2009; 11: 2218–2227.
[82] Gianolio D, Groppo E, Estephane J, Prestipino C, Nikitenko S, Zecchina A, Bordiga S, Taoufik M, Quadrelli EA, Basset JM, Lamberti C. Formation and reactivity of Cr(II) carbonyls hosted in polar and non polar supports. J. Phys.: Conf. Series 2009; 190.
[83] Estephane J, Groppo E, Vitillo JG, Damin A, Gianolio D, Lamberti C, Bordiga S, Quadrelli EA, Basset JM, Kervern G, Emsley L, Pintacuda G, Zecchina A. A multitechnique approach to spin-flips for $Cp_2Cr(II)$ chemistry in confined state. J. Phys. Chem. C 2010; 114: 4451–4458.
[84] Marra GL, Fitch AN, Zecchina A, Ricchiardi G, Salvalaggio M, Bordiga S, Lamberti C. Cation location in dehydrated Na-Rb-Y zeolite: An XRD and IR study. J. Phys. Chem. B 1997; 101: 10653–10660.
[85] Gordon KR, Warren KD. Spectroscopic and magnetic studies of the 3d metallocenes. Inorg. Chem. 1978; 17: 987–994.
[86] Ozin GA, Godbert J. Intrazeolite metallocenes: Chemistry, spectroscopy, and dynamics. J. Phys. Chem. 1989; 93: 878–893.
[87] Note that the scheme is not strictly rigorous. The electronic transitions should be expressed in terms of group theory once all the possible excited configurations have been taken into account. The arrows representing bands **I**, **II** and **III** have to be considered only as simplified to illustrate which electron is undergoing a transition and towards which type of molecular orbital. Notwithstanding the lack of a rigorous language, Figure 5.7b has the advantage of showing in an easy way and with a single scheme the origin of both the optical absorptions and the XANES features of $CrCp_2$ molecules. This scheme represents one of the few attempts in literature to match the information obtained with these two techniques.
[88] Warren KD. Ligand field theory of metal sandwich complexes. Configurations. Inorg. Chem. 1974; 13: 1243–1246.
[89] Muraoka PT, Byun D, Zink JI. Wavelength dependence of photooxidation vs photofragmentation of chromocene. J. Phys. Chem. A 2001; 105: 8665–8671.
[90] Brennan J, Cooper G, Green JC, Payne MP, Redfern CM. A study of valence level photoelectron cross-sections for the metallocenes of vanadium, chromium, cobalt and nickel from 20 to 100 eV. J. Electron Spectrosc. Relat. Phenom. 1993; 66: 101–115.
[91] Chetwynd-Talbot J, Grebenik P, Perutz RN. Photochemical reactions of M (ηt $C_5H_5)_2L_n$ (M = W, Mo, Cr, V) in low-temperature matrices. Detection of tungstenocene and molybdenocene. Inorg. Chem. 1982; 21: 3647–3657.
[92] Lavrentyev AA, Nikiforov IY, Dubeiko VA, Gabrelian BV, Rehr JJ. The use of the FEFF8 code to calculate the XANES and electron density of states of some sulfides. J. Synchrot. Radiat. 2001; 8: 288–290.

[93] Ankudinov AL, Rehr JJ. Development of XAFS theory. J. Synchrot. Radiat. 2003; 10: 366–368.
[94] Ruiz-Lopez MF, Loos M, Goulon J, Benfatto M, Natoli CR. Reinvestigation of the EXAFS and XANES spectra of ferrocene and nickelocene in the framework of the multiple scattering theory. Chem. Phys. 1988; 121: 419–437.
[95] Lamberti C, Bordiga S, Geobaldo F, Zecchina A, Otero Arean C. Stretching frequencies of cation CO adducts in alkali-metal exchanged zeolites – an elementary electrostatic approach. J. Chem. Phys. 1995; 103: 3158–3165.
[96] (a) Salassa L, Borfecchia E, Ruiu T, Garino C, Gianolio D, Gobetto R, Sadler PJ, Cammarata M, Wulff M, Lamberti C. Photo-induced pyridine substitution in *cis*-$[Ru(bpy)_2(py)_2]Cl_2$: a snapshot by time-resolved X-ray solution scattering. Inorg. Chem 2010; 29: 11240–11248. (b) Salassa L, Ruiu T, Garino C, Pizarro AM, Bardelli F, Gianolio D, Westendorf A, Bednarski PJ, Lamberti C, Gobetto R, Sadler PJ. EXAFS, DFT, light-induced nucleobase binding, and cytotoxicity of the photoactive complex *cis*-$[Ru(bpy)_2(CO)Cl]^+$. Organometallics 2010; 29: 6703–6710.
[97] Bonino F, Chavan S, Vitillo JG, Groppo E, Agostini G, Lamberti C, Dietzel PDC, Prestipino C, Bordiga S. Local structure of CPO-27-Ni metallorganic framework upon dehydration and coordination of NO. Chem. Mat. 2008; 20: 4957–4968.
[98] Salassa L, Garino C, Salassa G, Nervi C, Gobetto R, Lamberti C, Gianolio D, Bizzarri R, Sadler PJ. Ligand-selective photodissociation from $[Ru(bpy)(4AP)_4]^{2+}$: a spectroscopic and computational study. Inorg. Chem. 2009; 48: 1469–1481.
[99] Salassa L, Gianolio D, Garino C, Salassa G, Borfecchia E, Ruiu T, Nervi C, Gobetto R, Bizzarri R, Sadler PJ, Lamberti C. Structure of $[Ru(bpy)_n(AP)_{(6-2n)}]^{2+}$ homogeneous complexes: DFT calculation vs. EXAFS. J. Phys. Conf. Ser. 2009; 190: 012141.
[100] Cavka JH, Jakobsen S, Olsbye U, Guillou N, Lamberti C, Bordiga S, Lillerud KP. A new zirconium inorganic building brick forming metal organic frameworks with exceptional stability. J. Am. Chem. Soc. 2008; 130: 13850–13851.
[101] Valenzano L, Civalleri B, Chavan S, Bordiga S, Nilsen MH, Jakobsen S, Lillerud KP, Lamberti C. Disclosing the complex structure of UiO-66 Metal Organic Framework: A synergic combination of experiment and theory. Chem. Mat. 2011; 23: 1700–1718.
[102] Chavan S, Vitillo JG, Gianolio D, Zavorotynska O, Civalleri B, Jakobsen S, Nilsen MH, Valenzano L, Lamberti C, Lillerud KP, Bordiga S. H_2 storage in isostructural UiO-67 and UiO-66 MOFs. Phys. Chem. Chem. Phys. 2012; 14: 1614–1626.
[103] Kandiah M, Nilsen MH, Usseglio S, Jakobsen S, Olsbye U, Tilset M, Larabi C, Quadrelli EA, Bonino F, Lillerud KP. Synthesis and stability of tagged UiO-66 Zr-MOFs. Chem. Mater. 2010; 22: 6632–6640.
[104] Jakobsen J, Gianolio D, Wragg D, Nilsen MH, Emerich H, Bordiga S, Lamberti C, Olsbye U, Tilset M, Lillerud KP. Structural determination of a highly stable metal-organic framework with potentials in interim radioactive waste scavenging: Hf-UiO-66. Phys. Rev. B 2012; 86: art. n. 125429.
[105] Gianolio D, Vitillo JG, Civalleri B, Bordiga S, Olsbye U, Lillerud KP, Valenzano L, Lamberti C. Combined study of structural properties on metal-organic frameworks with same topology but different linkers or metal. J. Phys. Conf. Ser. 2013; 430: art. n. 012134.
[106] Milanesio M, Artioli G, Gualtieri AF, Palin L, Lamberti C. Template burning inside TS-1 and Fe-MFI molecular sieves: An in situ XRPD study. J. Am. Chem. Soc. 2003; 125: 14549–14558.
[107] Hafizovic J, Bjorgen M, Olsbye U, Dietzel PDC, Bordiga S, Prestipino C, Lamberti C, Lillerud KP. The inconsistency in adsorption properties and powder XRD data of MOF-5 is rational-

ized by framework interpenetration and the presence of organic and inorganic species in the nanocavities. J. Am. Chem. Soc. 2007; 129: 3612–3620.
[108] Agostini G, Lamberti C, Palin L, Milanesio M, Danilina N, Xu B, Janousch M, van Bokhoven JA. In situ XAS and XRPD parametric Rietveld refinement to understand dealumination of Y zeolite catalyst. J. Am. Chem. Soc. 2010; 132: 667–678.
[109] Dovesi R, Orlando R, Civalleri B, Roetti C, Saunders VR, Zicovich-Wilson CM. CRYSTAL: a computational tool for the ab initio study of the electronic properties of crystals. Z. Kristallogr. 2005; 220: 571–573.
[110] Martinez-Criado G, Borfecchia E, Mino L, Lamberti C. Micro and nano X-ray beams, in: Lamberti C, Agostini G (eds.), Characterization of Semiconductor Heterostructures and Nanostructures II, pp. 361–412, Amsterdam: Elsevier, 2013.
[111] Kazimirov A, Bilderback DH, Huang R, Sirenko A, Ougazzaden A. Microbeam high-resolution diffraction and X-ray standing wave methods applied to semiconductor structures. J. Phys. D-Appl. Phys. 2004; 37: L9–L12.
[112] Kazimirov A, Sirenko AA, Bilderback DH, Cai ZH, Lai B, Huang R, Ougazzaden A. Synchrotron high angular resolution microdiffraction analysis of selective area grown optoelectronic waveguide arrays. J. Phys. D-Appl. Phys. 2006; 39: 1422–1426.
[113] Mino L, Gianolio D, Agostini G, Piovano A, Truccato M, Agostino A, Cagliero S, Martinez-Criado G, Codato S, Lamberti C. Structural characterization of multi-quantum wells in electroabsorption-modulated lasers by using synchrotron radiation micrometer-beams. Adv. Mater. 2010; 22: 2050–2054.
[114] Cagliero S, Borfecchia E, Mino L, Calore L, Bertolotti F, Martinez-Criado G, Operti L, Agostino A, Truccato M, Badica P, Lamberti C. Insight into non-linearly shaped superconducting whiskers via a synchrotron nanoprobe. Superconduct. Sci. Technol. 2012; 25: 125002.
[115] Truccato M, Lamberti C, Prestipino C, Agostino A. Evidence of ion diffusion at room temperature in microcrystals of the $Bi_2Sr_2CaCu_2O_{8+\sigma}$ superconductor. Appl. Phys. Lett. 2005; 86: Art. n. 213116.
[116] Mino L, Gianolio D, Agostini G, Piovano A, Truccato M, Agostino A, Cagliero S, Martinez-Criado G, d'Acapito F, Codato S, Lamberti C. μm-EXAFS, μm-XRF, and μm-PL characterization of a multi-quantum-well electroabsorption modulated laser realized via selective area growth. Small 2011; 7: 930–938.
[117] Sirenko AA, Reynolds CL, Peticolas LJ, Ougazzaden A, Kazimirov A, Huang R, Fontes E, Bilderback D. Micro-X-ray fluorescence and micro-photoluminescence in InGaAsP and InGaAs layers obtained by selective area growth. J. Cryst. Growth 2003; 253: 38–45.
[118] Mino L, Agostino A, Codato S, Lamberti C. Study of epitaxial selective area growth $In_{1-x}Ga_xAs$ films by synchrotron μm-XRF mapping. J. Anal. At. Spectrom. 2010; 25: 831–836.
[119] Mino L, Agostino A, Codato S, Martinez-Criado G, Lamberti C. Effect of selective area growth mask width on multi-quantum-well electroabsorption modulated lasers investigated by synchrotron radiation X-ray microprobe. Nucl. Instr. Meth. Phys. Res. B 2012; 284: 6–9.
[120] Borfecchia E, Mino L, Gianolio D, Groppo C, Malaspina N, Martinez-Criado G, Sans JA, Poli S, Castelli D, Lamberti C. Iron oxidation state in garnet from a subduction setting: a micro-XANES and electron microprobe ("flank method") comparative study. J. Anal. At. Spectrom. 2012; 27: 1725–1733.
[121] Salvatore RA, Sahara RT, Bock MA, Libenzon I. Electroabsorption modulated laser for long transmission spans. IEEE J. Quantum Electron. 2002; 38: 464–476.

[122] Foti E, Fratta L, Ghiglieno F, Coriasso C, Cacciatore C, Rigo C, Agresti M, Vallone M, Codato S, Fornuto G, Fang R, Rosso M, Buccieri A, Valenti P. Optimisation of 10 Gbit/s InGaAsP electroabsorption modulator operating at high temperature. IEE Proc.-Optoelectron. 2004; 151: 103–108.
[123] Gibbon M, Stagg JP, Cureton CG, Thrush EJ, Jones CJ, Mallard RE, Pritchard RE, Collis N, Chew A. Selective-area low-pressure MOCVD of GaInAsP and related materials on planar InP substrates. Semicond. Sci. Technol. 1993; 8: 998–1010.
[124] Coltrin ME, Mitchell CC. Mass transport and kinetic limitations in MOCVD selective-area growth. J. Cryst. Growth 2003; 254: 35–45.
[125] Romanato F, De Salvador D, Berti M, Drigo A, Natali M, Tormen M, Rossetto G, Pascarelli S, Boscherini F, Lamberti C, Mobilio S. Bond-length variation in $In_xGa_{1-x}As/InP$ strained epitaxial layers. Phys. Rev. B 1998; 57: 14619–14622.
[126] d'Acapito F. Nearest-neighbor distances in strained thin films of random pseudobinary semiconductor alloys: A calculational methodology. J. Appl. Phys. 2004; 96: 369–373.
[127] Borfecchia E, Garino C, Salassa L, Ruiu T, Gianolio D, Zhang XY, Attenkofer K, Chen LX, Gobetto R, Sadler PJ, Lamberti C. X-ray transient absorption structural characterization of the ^{3}MLCT triplet excited state of cis-$[Ru(bpy)_2(py)_2]^{2+}$. Dalton Trans. 2013; 42: 6564–6571.
[128] Gianolio D, Borfecchia E, Garino C, Ruiu T, Lamberti C, Salassa L. The use of differential EXAFS analysis for the determination of small structural differences between two closely-related ruthenium complexes. J. Phys.: Conf. Series 2013; 430: art. n. 012125.

Cinzia Giannini, Davide Altamura, Brunella Maria Aresta, Teresa Sibillano, Dritan Siliqi, and Liberato De Caro

6 Lens-less scanning X-ray microscopy with SAXS and WAXS contrast

Independently of the type of source (visible light, X-rays, electrons) a microscope generally allows to produce enlarged images of the illuminated sample. In the most popular (optical or electron) microscopy techniques, this results in the direct imaging of the investigated sample or a portion of it. As a consequence, all the observed details in the image are assumed to represent actual structures of the sample. In contrast, diffraction-based microscopy provides indirect (Fourier space) images of the sample, so that retrieving sample structure is in general not straightforward, but important, highly-detailed information (*e.g.*, atomic structure) can be obtained, depending on radiation wavelength.

In this chapter we will briefly introduce a quite recent microscopy technique, based on X-ray diffraction, known as scanning X-ray microscopy with small angle (SAXS) and wide angle (WAXS) contrast. We will describe the basic principles behind these techniques and show a couple of applications to the nano-bio world. Indeed, the high penetration depth, high achievable resolution, and the basically nondestructive character of X-rays are among the main advantages which make X-ray techniques extremely useful for the study of nanostructures.

It is suggested to start the present chapter after reading the chapter of this book entitled *Single crystal and powder XRD techniques: an overview* by A. Altomare *at al.* which already contains several useful and basic concepts of crystallography and powder diffraction analysis.

6.1 Introduction

Continuous progress has been registered in the last decades in preparing materials at the nanoscale, as well as in fabricating nanoscopic hierarchical structures made by a blend of synthetic inorganic materials and biological systems [1]. The material properties vary as a function of atomic structure, size, size distribution, and shape of the nanodomains. Each of the aforesaid features can play an important role in producing unique physical-chemical properties of relevant interest in a countless number of applications: catalysis, magnetic resonance imaging, data storage, optics, optoelectronics, biotechnology/biomedicine, industrial and environmental applications [2]. Various synthetic approaches have been optimized over the years and are nowadays able to produce finely controlled nanocrystals (NCs) of metals, oxides and semiconductors, and to tailor nanoscience and nanotechnology towards more complex objects (*e.g.*, hybrid NCs, assembled nanocrystals that can be designed, on purpose, for multitasking applications).

Despite the very exciting outlook offered by the synthesis routes, an exhaustive structural and microstructural characterization of the NCs, together with the spatial organization of the nanostructures, is difficult to achieve, limiting the deep understanding of the unique properties, and the development of new functional materials. The lack of sufficient knowledge about the atomic structure of NCs is often related to the complexity of the problem, due for example to a very small size of the NCs (only a few nanometers), to their size and/or shape polydispersity, and mixing of different crystalline phases. Specialized structural and morphological characterization tools are needed to quantitatively extract structure, size, size-dependent strain and shape distributions of selected nanocrystalline materials but, at the same time, innovative and complete data collection techniques are necessary to use the developed characterization tools in the most synergic and efficient way.

6.2 X-ray microscopes

There are two main principles of microscopes to be distinguished:
- full-field microscopes;
- scanning microscopes.

In full-field microscopes the whole field of view is imaged to a detector plane at the same time. In scanning microscopes the sample is illuminated with a bright well-focused spot scanning over the sample. The detector then measures the total intensity over time coming from the currently illuminated spot on the sample and the image is extracted from these data when the scanning process is finished.

Scanning microscopes (Fig. 6.1) consist of a source, with focusing illumination optics, a motorized sample holder, imaging objective lens or optics, and a detector to collect the resulting image [3]. The intensity of the source determines the time needed to take an image with a good signal to noise ratio. The focusing illumination optics has to be able to concentrate the incoming beam delivered at the sample, to guarantee an optimal overall imaging quality of the sample. Nowadays several different X-ray optics are available [4–8], working in reflection, refraction or diffraction mode, and having different performances in terms of flux, achromaticity, coherence, spot size *etc.* (see Fig. 6.2 for comparison).

A typical set-up is composed as follows: the sample is fixed in a sample holder; the holder (or the stage it is mounted on) is motorized for precise positioning of the sample in the beam, allowing to choose a section of interest, as well as to adjust the magnification of the system. The imaging lens has to be suited for the beam energy used and its acceptance angle needs to be large enough to image the desired field of view. In this way the objective lens images each point of the sample within the field of view to a corresponding point in the image plane. The detector placed in the image

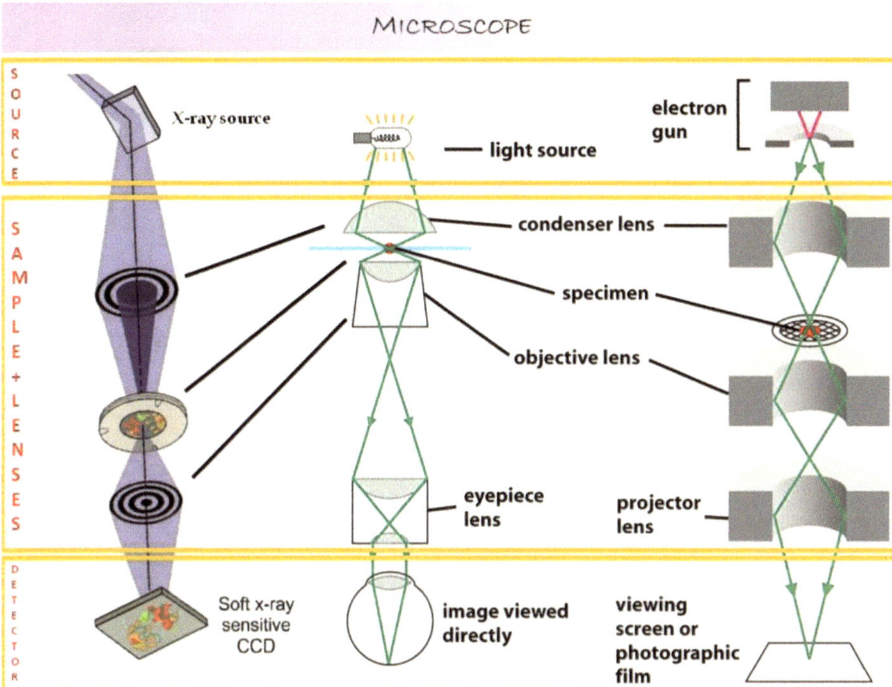

Fig. 6.1: Microscopes with different type of sources.

plane registers the intensity distribution scattered by the sample, and returns an image of the sample.

As for other sources, also in the case of X-ray microscopes, there is an objective lens which recombines the waves scattered by the sample with correct phases to give the image. Unfortunately, chromatic and spherical aberrations of the lens severely affect the ultimate available spatial resolution in the image. An alternative to objective lenses is given by a lens-less set-up as schematically shown on the bottom part of Figure 6.3, in which the X-ray beam scattered by the sample freely propagates until it reaches the 2D detector. The resolution in both imaging methods is given by $\frac{\sin(\Theta)}{\lambda}$, with λ being the wavelength and Θ the maximum scattering angle accepted by the lens or by the detector (numerical aperture). In the case of lens-less imaging, the resolution is practically limited by the wavelength, by the detector angular acceptance, by photon noise, mechanical precision and stability, and by the sample radiation damage [9]. Note that, in the case of a lens-less microscope without an objective lens, the registered intensity does not directly give the image of the sample. The detected intensity contains structural and morphological information of the object. This information has to be decoded somehow from the scattered data.

		Characteristics	
Reflective	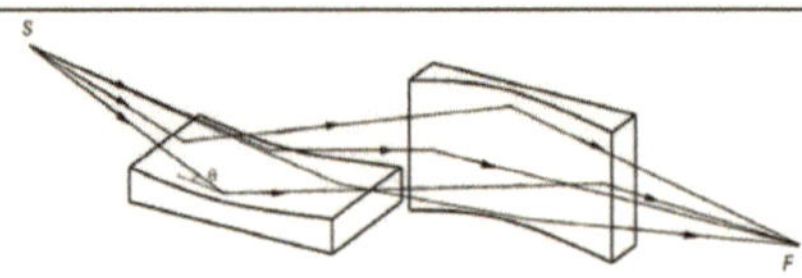	E	<30KeV
		ρ	25nm @15KeV
		α	YES
		ϕ	+++
		E	<20KeV
		ρ	50nm
		α	YES
		ϕ	---
	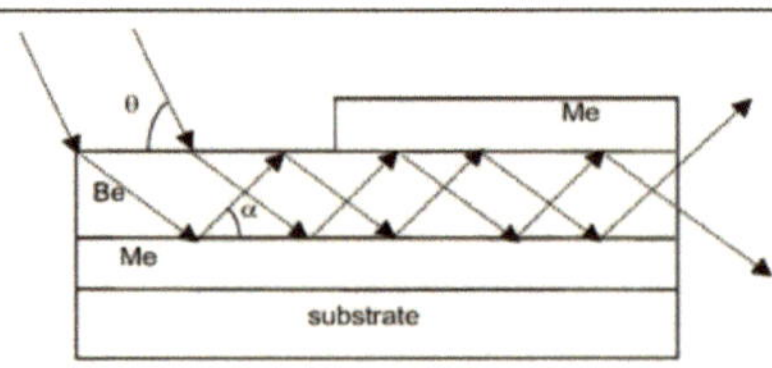	E	<20KeV
		ρ	45x25nm^2
		α	NO
		ϕ	---
Diffractive	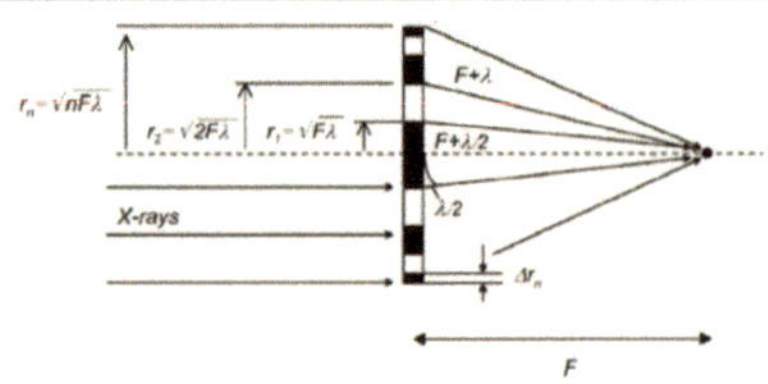	E	<30KeV
		ρ	30nm @20KeV
		α	NO
		ϕ	+++
Refractive	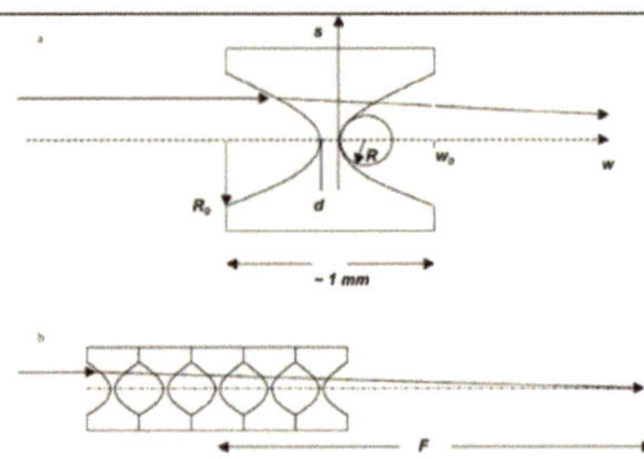	E	<1MeV
		ρ	50nm @20KeV
		α	NO
		ϕ	+

Fig. 6.2: X-ray optics: E = working energy, ρ = resolution; α = achromaticity; Φ = flux (adapted from [4]).

In the following we will deal with a scanning lens-less X-ray microscope having a small-angle and wide-angle scattering contrast (SAXS and WAXS).

In the realm of the most known X-ray imaging techniques (Fig. 6.4), scanning lens-less microscopes with SAXS and WAXS contrast are fundamentally used to inspect biological tissues and natural materials with hierarchic orders, such as bone, wood, skin, teeth, horns, but also artificial nanostructured materials.

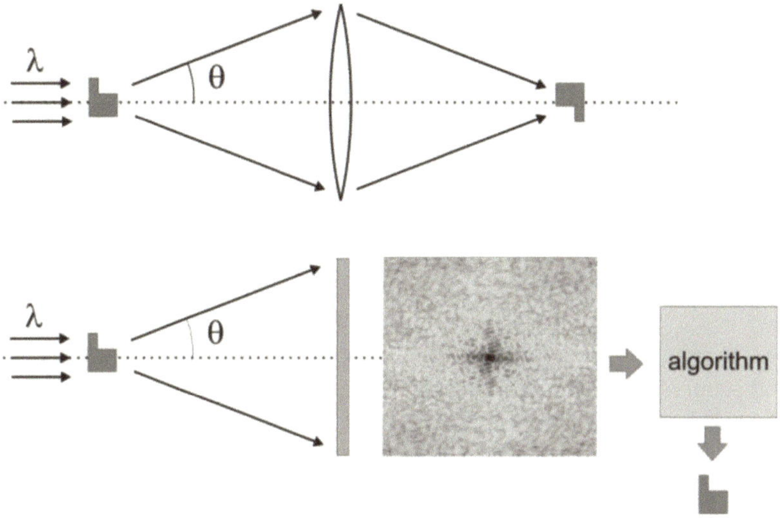

Fig. 6.3: Top: sketch of a conventional full-field imaging setup with a lens. Bottom: schematic lens-less imaging set-up. Reproduced from [9].

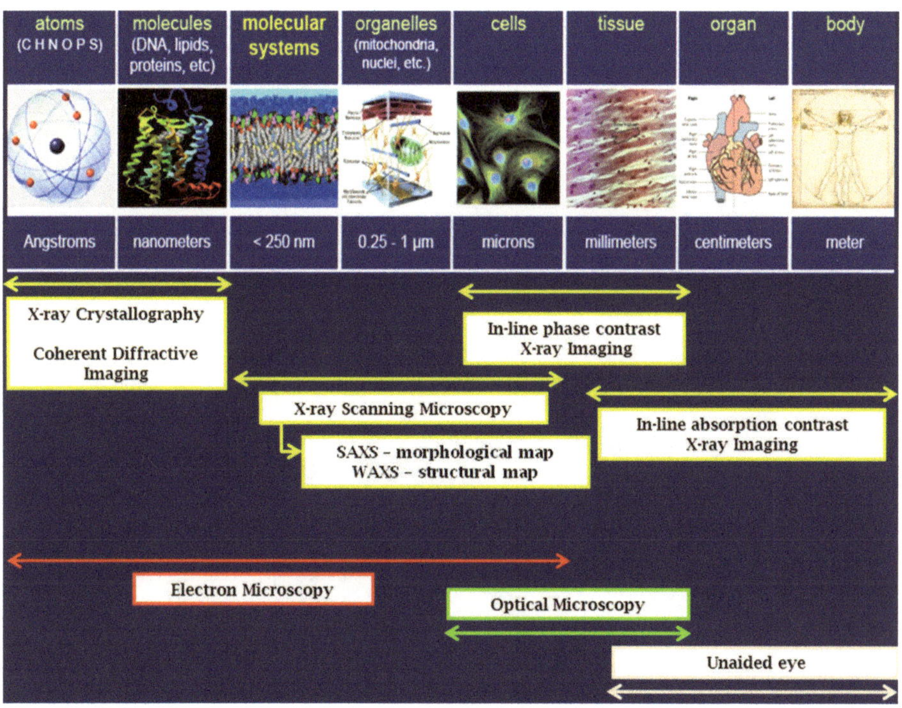

Fig. 6.4: X-ray imaging techniques. Courtesy of Dr. Frank Chuang.

New generation X-ray laboratory sources are available with brightness exceeding some second generation bending magnet synchrotron beamlines. Coupling such a microsource to suitable X-ray optical devices allows to produce X-ray beams of micrometer size and high intensity, thus enabling X-ray diffraction and X-ray imaging experiments from spots also in laboratories. In the past, such small spots were only the prerogative of synchrotron radiation sources. Indeed, in every European synchrotron radiation center, dedicated microfocus beamlines can routinely provide a microbeam with sizes down to 1 μm (ESRF, BESSY II, Diamond, PETRA III, SOLEIL, are among the most representative examples). At synchrotrons, the big effort under way nowadays is to push the X-ray source + focusing optics systems down to nanometric values, to improve the detection technology and to combine spectroscopic, structural and elemental techniques into the same inspection tool. Although the best performances can be obtained with synchrotron radiation, unrivalled intensities of 1.6×10^{11} photons $mm^{-2} sec^{-1}$ can be obtained with these new Rigaku FR-E+ SuperBright synchrotron-class generators. With these novel rotating anode synchrotron-class generators for microfocus X-ray high brilliance sources, all the high performing structural techniques are no longer exclusively a prerogative of synchrotron sources which are normally overbooked and with access conditional on the submission of a proposal and a screening process by a referee committee.

Figure 6.5 shows the scanning microscope located at the CNR-IC Institute in Bari [10]. The laboratory (XMI-L@b) is equipped with an Fr-E+ SuperBright rotating copper

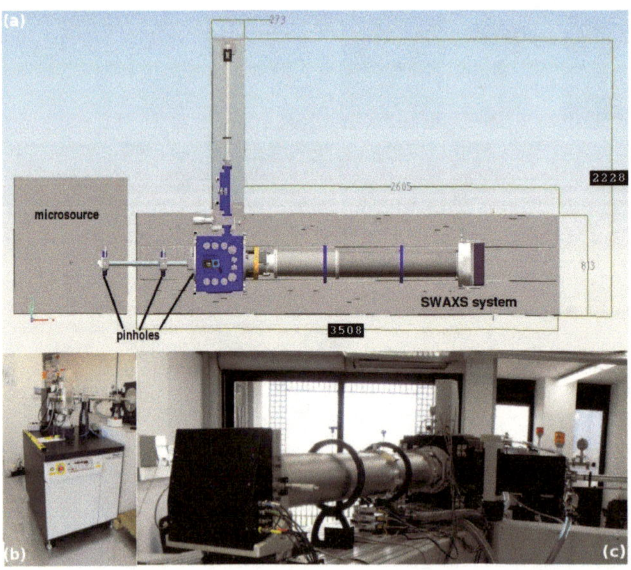

Fig. 6.5: (a) Scheme of the XMI-L@B; (b) SuperBright rotating copper anode microsource (45 kV/55 mA; Cu-Kα, λ = 0.15405 nm); (c) SAXS/WAXS (SWAXS) three-pinhole camera. Reproduced with permission of the International Union of Crystallography (http://journals.iucr.org/) from [10].

anode microsource (45 kV/55 mA; Cu-Kα, $\lambda = 0.15405$ nm, $9 * 10^6$ photons/sec, 2475 W) shown in Figure 6.5b, coupled through a focusing multilayer optics (Confocal Max-Flux (CMF® 15-105)) to a SAXS/WAXS (SWAXS) three-pinhole camera equipped for scanning microscopy (Fig. 6.5c). The Fr-E+ SuperBright microsource is quite unique in terms of brilliance as a laboratory source, with its flux being comparable to a bending magnet synchrotron light source. When working as a scanning microscope, a stage is used to move the sample in the beam with two detectors measuring, also simultaneously, the intensity scattered by the sample at small and wide angles. The scanning stage moves the sample through the beam, line by line, until the whole field of view is covered. The required step size of the stage depends on the desired image resolution. As a guideline, the step size should be about half the beam spot size at the sample to reach maximum possible image quality (here 70 μm diameter). A Triton™ 20 gas-filled proportional counter (1024 × 1024 array, 195 μm pixel size) is adopted for SAXS acquisition, whereas an image plate (IP) detector (250 × 160 mm^2 in size, with 50 or 100 μm effective pixel size depending on binning), with off-line RAXIA reader, is employed to collect WAXS. A lock and load feed system is mounted on the side of the main SAXS chamber to insert and remove the image plate without breaking the vacuum.

The following section illustrates what information is encoded in SAXS and WAXS patterns.

6.3 Small-angle and wide-angle scattering contrast (SAXS and WAXS)

As already explained in Chapter 4 of this book entitled *Single crystal and powder XRD techniques: an overview* by A. Altomare *at al.*, the interaction of X-rays with matter is basically restricted to scattering with electrons, because of their much smaller mass with respect to protons.

Classically, X-ray scattering from electrons can be understood as an elastic scattering from free electrons oscillating in the electric field of the electromagnetic wave. The electric field of the incident wave (photon), $e^{i(\boldsymbol{k}\cdot\boldsymbol{r}_n-\omega t)}$, accelerates the electrons who irradiate electromagnetic waves at the same frequency as the incident wave. A simple treatment of this classical analysis is given by the Thomson cross section for one electron:

$$\frac{d\sigma}{d\Omega} = \frac{r_e^2}{2}(1+\cos^2 2\theta),$$

where 2θ is the scattering angle, $r_e^2 = \frac{e^2}{4\pi mc^2\epsilon_0}$ is the classical radius of the electron ($2.82\cdot 10^{-15}$ m) and X-rays are assumed unpolarized (as is the case for X-rays produced by conventional tubes). From this formula it is readily seen that the Thomson cross section depends on the inverse of the mass of the scatterer, and this fact implies that the scattering from the atoms is mainly due to the scattering from the electrons, much lighter than protons.

Scattering from atoms is mainly due to scattering from individual electrons, but the scattering amplitudes do not simply add up to make the total scattering intensity of the atom. The phase differences between the waves complicate calculations. The only case in which the amplitudes simply add up is when the scattering angle is zero. Scattering factor for atoms is defined as

$$A = \frac{\text{the scattering amplitude of all electrons in the whole atom}}{\text{the scattering amplitude of one electron}}$$

The scattering factor has its maximum in the forward direction, where it equals the number of electrons and it decreases monotonically as a function of the scattering angle. The values of atomic scattering factors for different atoms and angles are tabulated in the International Tables of Crystallography (http://it.iucr.org/).

In Figure 6.6 the Cu atomic scattering factor behavior is represented, which shows the common features of all atomic scattering factor functions:

- for $\theta = 0$, f = the number of electrons in the atom;
- f is monotonically decreasing;
- f shows the "fundamental" behavior of scattering from disordered matter.

Atomic scattering factors account only for single coherent (*i.e.*, Thomson) scattering processes. Incoherent (*i.e.*, Compton) scattering comes out as a background signal.

Scattering factors also depend on the energy of the incoming photon. A more general and convenient expression for the atomic scattering factor is

$$f(k, E) = f_0(k) + f'(E) + if''(E),$$

where the energy dependence is expressed by the second and third terms on the right-hand side, which are only significant around the absorption edges of the sample material. These terms can be calculated from the measured absorption spectrum by the

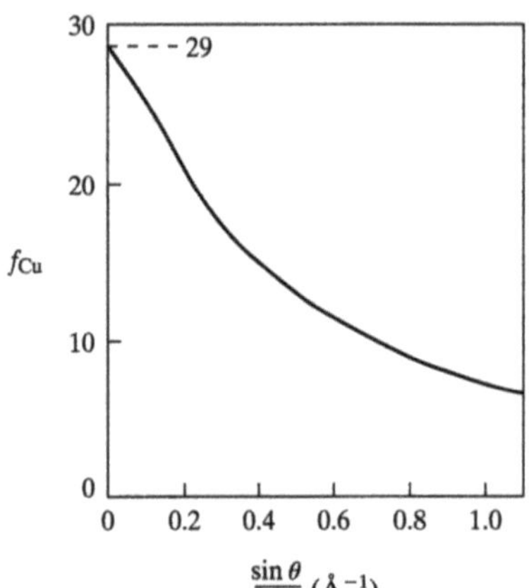

Fig. 6.6: Atomic scattering factor of Cu (adapted from [11]).

use of the following formulas:

$$\mu(E) = 2N\lambda r_e f''(E)$$

and

$$f'(E) = \frac{2}{\pi} P \int_0^\infty \frac{E' f''(E')}{E'^2 - E^2} dE',$$

where μ is the absorption coefficient, N is the atomic number and P denotes the Cauchy principle value of the integral [11].

It is not only in the sub-atomic electron density distributions that the phase differences of the waves have significance. Actually, the electron density distribution differences at the atomic and nanometer scales, producing significantly larger phase shifts in the scattered waves, give rise to the formation of WAXS and SAXS patterns, respectively. The electron density function $\rho(r)$ contains all of the information that is needed in practice to fully describe the material structure, namely *the high-resolution* material structure (*i.e.* the atomic configuration) and *the low-resolution* features of the structure (*i.e.* the morphology). The scattering amplitude and intensity are related to the elastic interaction between a narrow monochromatic X-ray beam – with a wavelength λ – and an arbitrary material, with electron density function $\rho(r)$. They are functions of the scattering vector $q = \frac{4\pi}{\lambda} \sin\theta$ (here θ is half the scattering angle) that identifies the length scale d of the scattering objects probed in the experiment through $q = \frac{2\pi}{d}$. At small scattering angles (2θ less than a few degrees) typical length scales vary from 1 to 100 nm (SAXS regime). At wide scattering angles (2θ between 5 and 100 deg) typical length scales are in the range of inter-atomic distances (WAXS regime). Therefore, SAXS will provide information on *the low-resolution* features of the structure, *i.e.* morphology of the particle (external size/shape) whereas WAXS provides information on *the high-resolution* material structure, *i.e.* the atomic scale order, or lattice periodicity. Ideally this information should be available within the same experiment, without removing the sample, as is possible at the XMI-L@B with the simultaneous collection of the SAXS and WAXS signals.

In crystals, atoms are organized in regular 1D, 2D or 3D lattices according to specific symmetries. Crystals organize in lattices which can be characterized by a unit cell which is totally described by its edges (a, b, c) and angles (α, β, γ) and can assume only few possible geometries (cubic, tetragonal, orthorhombic, trigonal-hexagonal, monoclinic and triclinic).

If the scattering factors and positions of the atoms in the unit cell are known, the total scattering factor of the unit cell can be calculated as

$$F(k) = \sum_{n=1}^{N} f_n e^{i\mathbf{k}\cdot\mathbf{r}_n},$$

where f_n are the atomic scattering factors of the individual atoms and $\mathbf{r}_n = x_n\mathbf{a} + y_n\mathbf{b} + z_n\mathbf{c}$ are the atomic positions in the lattice. The scattering amplitude of the unit cell is called structure factor. The total scattering amplitude of the crystal is

$$A = F \sum_{l=0}^{L-1} e^{i\mathbf{k}\cdot l\mathbf{a}} \sum_{m=0}^{M-1} e^{i\mathbf{k}\cdot m\mathbf{b}} \sum_{n=0}^{N-1} e^{i\mathbf{k}\cdot n\mathbf{c}}$$

where La, Mb, Nc represent the sizes of the 3D lattice along the directions of the base vectors $\boldsymbol{a}$, $\boldsymbol{b}$, $\boldsymbol{c}$.

In order to compute the scattering power of an assembly of randomly oriented nanocrystals, one has to average over all orientations. If the atoms are assumed point-like and their atomic scattering factors are constant with the scattering angle, it follows that

$$I(k) = \left\langle \sum_{i=1}^{N} f_i e^{ik\cdot r_i} \right\rangle = \sum_{i=1}^{M} \sum_{j=1}^{N} f_i f_j \int_0^{\pi} \int_0^{2\pi} d\theta d\varphi e^{ik\cdot(r_i - r_j)} = \sum_{i=1}^{M} \sum_{j=1}^{N} f_i f_j \frac{\sin kr_{ij}}{kr_{ij}}.$$

This is the simplest version of the Debye function (DF), that has been recently used to approach the analysis of WAXS data [12–14], but it can be used as well for SAXS data analysis [15]. The main advantage of using the Debye formula when investigating short-range ordered materials such as nanocrystalline compounds is the ability to simultaneously model both Bragg (if any) and diffuse sample scattering, all arising from the distribution of the interatomic distances within the sample. The Debye formulation is indeed independent of any lattice periodicity or crystallographic order, and is therefore the preferred approach if a quantitative analysis of domain size, unit cell distortion and structure is needed for a nanocrystalline material.

For extremely small nanocrystals, surface atoms can reach extremely large percentages (50–75 %), as shown in Figure 6.7 [16, 17]. This fraction of atoms sees a different environment from that seen by the interior atoms. Indeed, the process of minimization of the surface energy leads to important strain contributions and/or surface atoms lattice reconstruction that significantly affect the overall structure and the physical properties of the nanocrystals [2]. As a consequence, the precise determination of the nanocrystal structure, both the medium-range order and/or the existence of local distortion, is a fundamental issue.

The DF method allows to extract size, strain and structure distributions from the powder diffraction pattern collected on a nanomaterial. Size-dependent quantities like the lattice parameters and Debye–Waller factors can be derived as well [2].

As an alternative to the DF method, a more traditional whole profile fitting analysis of the pattern can be adopted, *i.e.* the Rietveld approach [18], modeling a shape anisotropy [19]. This alternative is less time consuming but still accurate for nanocrystals larger than 10–15 nm.

Several metallic (Au), oxide (Fe_2O_3, Fe_3O_4, TiO_2) and semiconductor (CdSe, CdTe, PbS) nanocrystals, just to cite a few, have been studied with both methods [2 and reference therein].

Full-shell Clusters	Total Number of Atoms	Surface Atoms (%)
1.Shell	13	92
2.Shells	55	76
3.Shells	147	63
4.Shells	309	52
5.Shells	561	45
7.Shells	1415	35

(a)

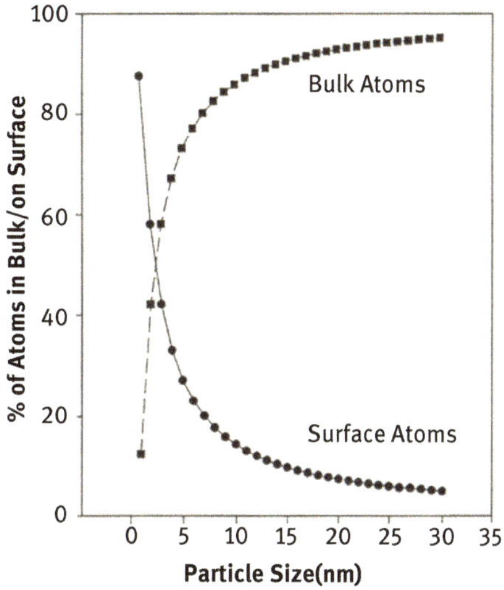

(b)

Fig. 6.7: Number of total and surface atoms for increasing nanocrystal size. Reproduced with permission from [16, 17].

It is worth repeating a peculiar difference which distinguishes SAXS from WAXS information. The SAXS signal carries morphological information (nano-object external shape) of the investigated nano-object, independently of its crystallinity. In other words, the nano-object can be studied with SAXS provided that its visibility (electron density contrast between nanoobject and embedding matrix) allows it to be detected. WAXS provides information on the structure within the crystalline domain. See Figure 6.8 for comparison.

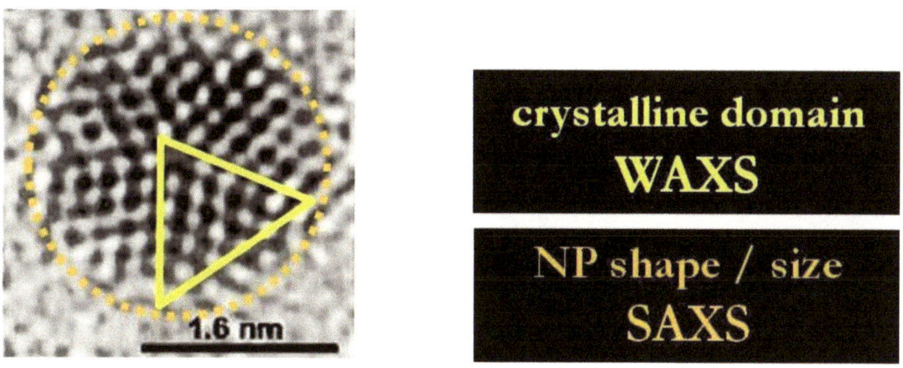

Fig. 6.8: Multiple domain gold nanocrystals, as inspected by high resolution transmission electron microscopy. Reproduced with permission from [20].

Figures 6.9 and 6.10 show the typical information encoded in a WAXS and SAXS pattern, respectively.

A WAXS pattern (Fig. 6.9) contains information on the nanocrystal's structure (and structure abundance), on the domain size and shape, on unit cell deformation and on the presence of defects. SAXS allows to measure the average morphology of randomly distributed and diluted nanocrystals (gas-type structures); nanocrystals with short-range spatial correlation (liquid-like structures); periodically arranged nanocrystals (crystal-like structures), as exemplified in Figure 6.10.

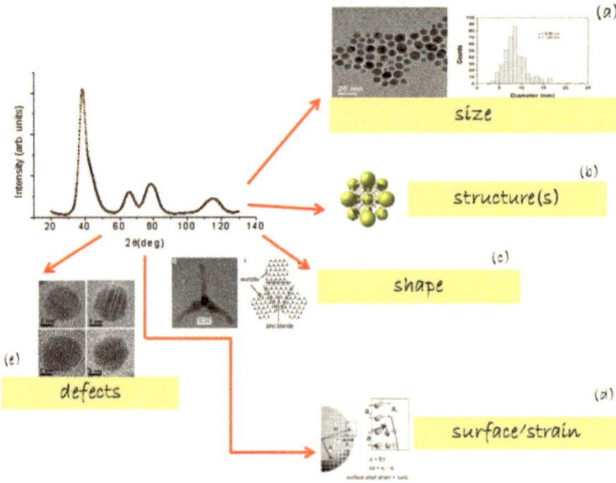

Fig. 6.9: WAXS – contribution encoded in a typical diffraction pattern (structure, size and shape of the crystalline domain, unit cell deformation induced by surface effects, defects). Reproduced with permission of Elsevier from [2].

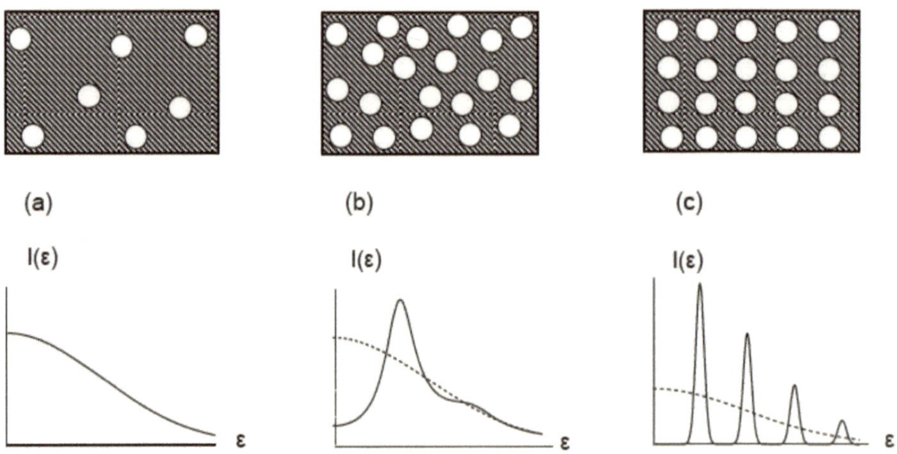

Fig. 6.10: schematic masks and corresponding light scattering intensity for (a) a gas-like (dilute) set of nanocrystals, (b) a liquid-like (concentrated) set of nanocrystals with short-range spatial order, and (c) a crystal-like set of nanocrystals with long range spatial order. Courtesy of Prof. AF Craievich [21].

6.4 Applications

Two examples of applications of scanning microscopy with SAXS/WAXS contrast in the research field of nanomedicine and biomaterials are described here.

The first example deals with studies of *bone biopsies* collected from patients affected by different pathologies.

Bone is a natural material with a hierarchical structure made of osteons with circular arrangements of differently oriented collagen fibers, each collagen fiber consisting of bundles of collagen fibrils [22–27]. The collagen fibril consists of a staggered arrangement of collagen molecules with embedded hydroxiapatite (HA) mineral nanocrystals. Interfibrillar mineralization of the HA nanocrystals typically occurs into the gaps between the collagen fibrils, which work as a natural template for the HA nanocrystal deposition, inducing a preferred orientation of the nanocrystals. Additionally, extrafibrillar mineralization may occur, whose fingerprint is the presence of randomly oriented nanocrystals (see Figure 6.11).

Scanning small- and wide-angle X-ray scattering (scanning SWAXS) and circularly polarized light (CPL) microscopy experiments were performed on healthy and pathologic (dwarfism syndrome and Paget's disease) human bone sections [26]. Paget's disease is also called osteitis deformans. It is a chronic disorder with alterations of the bone remodeling equilibrium and accelerated deposition of not well-organized lamellar bone, in particular for trabecular bone. Dwarfism is a medical condition that results in the underdevelopment of the body. It might be caused by genetic factors, nutritional, or hormone deficiencies.

The SAXS and WAXS raw data, acquired by using a scanning microscope, were transformed, *via* crystallographic tools, into quantitative radiographic images and compared with CPL images in order to map local changes, at an osteon detail level, of the interfibrillar and extrafibrillar HA nanocrystals within the collagen fibril (see Fig. 6.12, 6.13, and 6.14). The fundamental question for many studies performed in the last decades on bone pathologies is how is it possible that a mutation at the molecular/nanoscale level could lead to the failure of the macroscopic tissue. Indeed, answering this question is not trivial, but novel microscopic and nanoscopic methods are available to offer possible additional data.

With the present methods we shed light onto the geometry, orientation and architecture of the interfibrillar and extrafibrillar mineral nanocrystals within the mineralized collagen fibril, all indirect determinants of the bone strength/fragility and flexibility, by inspecting for the first time with these techniques dwarfism syndrome and Paget's disease biopsies.

The CPL (a) image is the map of the collagen fiber orientation within the osteons, while the SWAXS information (b, c, d) refers to the mineral counterpart. Precisely, panel (a) of Figures 6.12, 6.13, and 6.14 refers to the CPL microscopy, panel (b) to the scanning WAXS and panels (c) and (d) to the scanning SAXS microscopy. The red color in (b) indicates the presence of HA nanocrystals uniaxially oriented along the

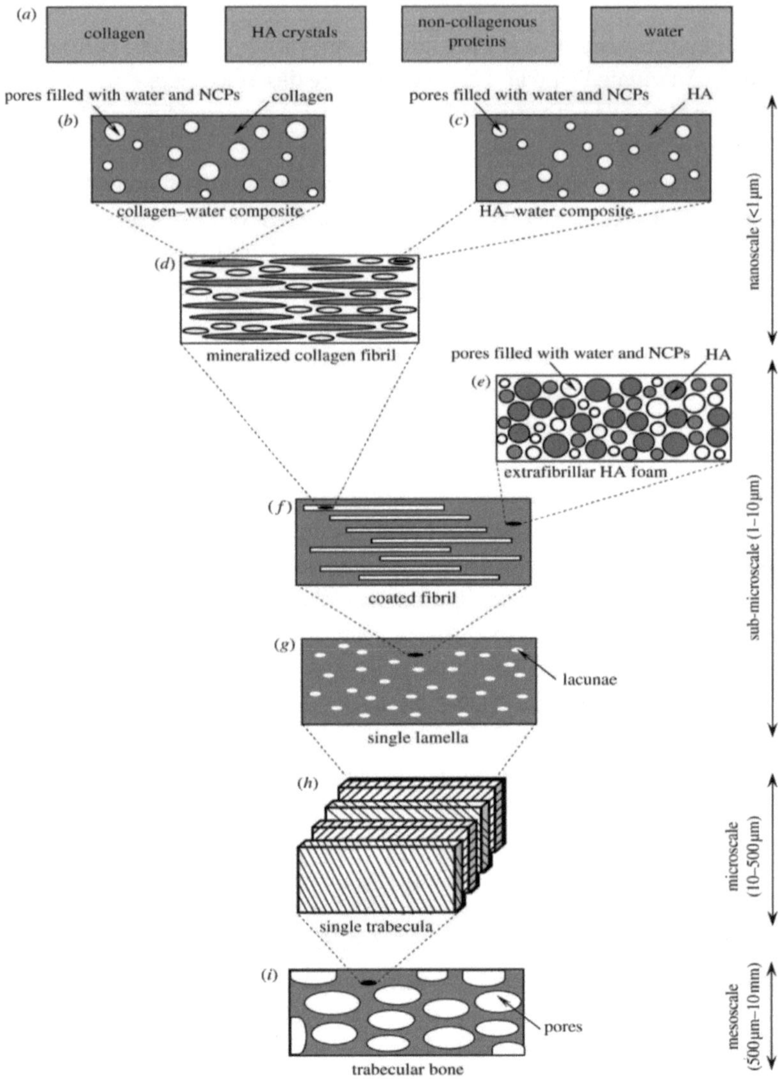

Fig. 6.11: A schematic showing all of the successive steps taken in the modeling of trabecular bone: (a) basic constituents of bone, (b) a collagen-water composite formed by cross-linked collagen matrix containing some pores filled with water and NCPs, (c) an HA-water composite built up by an HA crystal containing some intercrystalline pores filled with water and NCPs, (d) a mineralized collagen fibril made up of collagen-water and interfibrillar HA-water minerals, (e) an extra-fibrillar HA foam made up of HA minerals and some pores between them hosting water and NCPs, (f) a coated fibril formed through the interaction of mineralized collagen fibrils and extrafibrillar HA foam, (g) a single lamella built up from the coated fibrils matrix perforated by lacunar cavities, (h) several lamellae with different fibril orientation stacked together to form a single trabecula, and (i) a trabecular bone as a porous network of trabeculae. Reproduced with permission from [27].

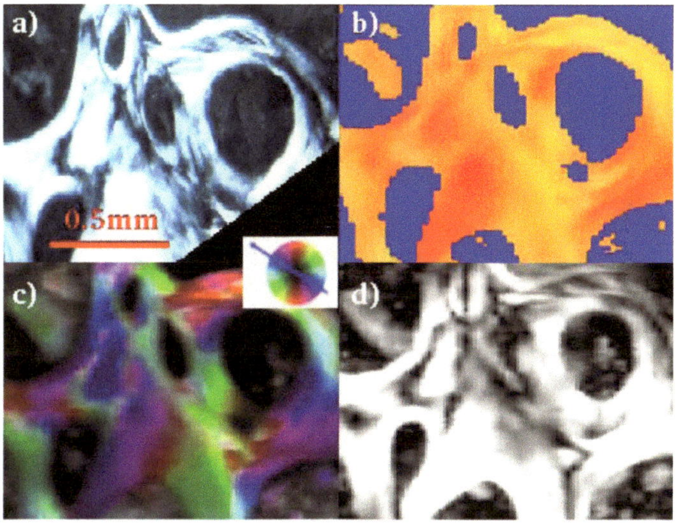

Fig. 6.12: Microscopies of the healthy biopsy. Reproduced with permission from [26].

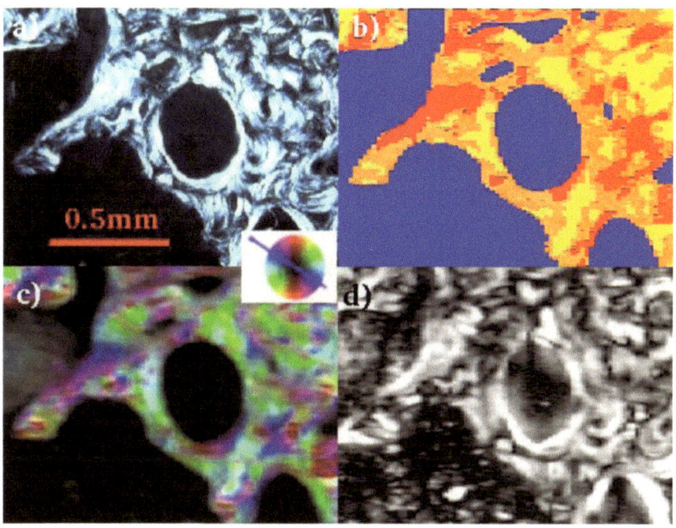

Fig. 6.13: Microscopies of the biopsy affected by Paget's disease. Reproduced with permission from [26].

c-axis as in the case of interfibrillar bone; the yellow color refers to the distribution of the extra-fibrillar bone mineralization but also to randomly oriented microfibrils (internally mineralized). The elongation direction and orientation degree of the HA nanocrystals are represented in the panels (c) and (d), respectively: the higher the degree of orientation, the brighter the intensity in panel (d).

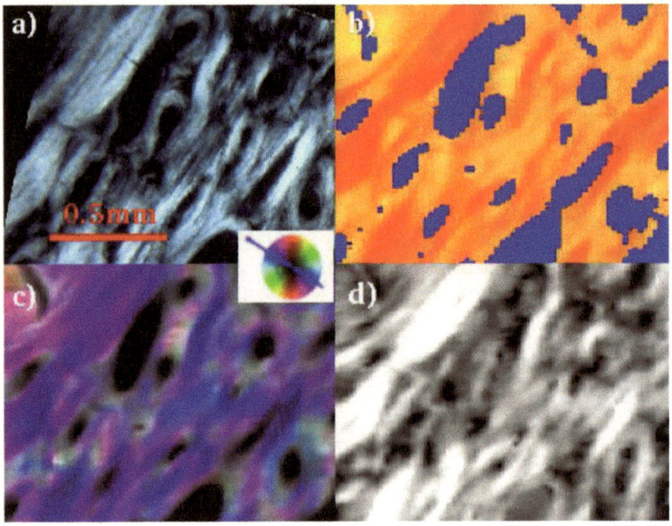

Fig. 6.14: Microscopies of the biopsy affected by dwarfism. Reproduced with permission from [26].

The SWAXS microscopies clearly indicate that the dwarfism disease biopsy has abnormally large areas with mono-oriented aligned interfibrillar mineral nanocrystals, with respect to the other biopsies, as SAXS data are practically dominated by a single orientation. In addition, the higher and lower orientation degree was measured by SAXS for the dwarfism and Paget's disease biopsies, respectively. The Paget's case shows more fragmented and disorganized structures. The age difference among patients, namely 27 for the dwarfism disease biopsy and more than 80 years for the others, may play a role.

The second example deals with studies of *type I collagen* collected from rat tail tendon and dried.

Collagen is the most abundant protein in the human body. Among the over 25 types of collagen known, fibril-forming collagen is the main component in many tissues such as tendons, cartilage and bone [28]. Fibril-forming collagen is characterized by a hierarchical assembly of substructures. In this hierarchical arrangement, collagen molecules consisting of three polypeptide chains assemble into fibrils with diameters in the range of 10–500 nm and the fibrils further assemble into fibers [29–30]. The main function of fibril-forming collagen is to provide the structural framework and the strength of tissues [29].

Collagen has been widely used in the manufacturing of cosmetics, glue and gelatin. In the past 30 years, collagen-based biomaterials have been developed for a number of medical and tissue engineering applications like, for example, heart valves, artificial skin substitutes and scaffolds for tissue engineering and for drug delivery applications [31–33].

In tendons, collagen molecules are triple helices with a length of approximately 300 nm, assembled into fibrils with a periodicity of $\sim$ 64–67 nm along the fiber [34–35]. This strong anisotropic order of the structure can be clearly inspected from the typical diffraction pattern of a SAXS image. The experimental set-up is rather simple: the orientated fiber is placed in a collimated X-ray beam at right angles to the beam and the “fiber diffraction pattern” is recorded on a detector placed a few cm or m away from the fiber, depending on the particular interplanar distances to be measured (1–2 m are needed to access nanometric distances). The example in Figure 6.15 belongs to the cases of SAXS from undiluted crystals and, more precisely, SAXS from 1D periodically oriented nanocrystals (Fig. 6.10c). Figure 6.15 shows the way the fiber has to be oriented with respect to the X-ray beam (panel (a)), the typical SAXS 2D image and corresponding azimuthally integrated 1D pattern (panels (b) and (c)), a map of the previous information across a $1.2 \times 0.8\,\mathrm{mm}^2$ area, and the fiber orientation map converted in colors, as indicated in the color wheel.

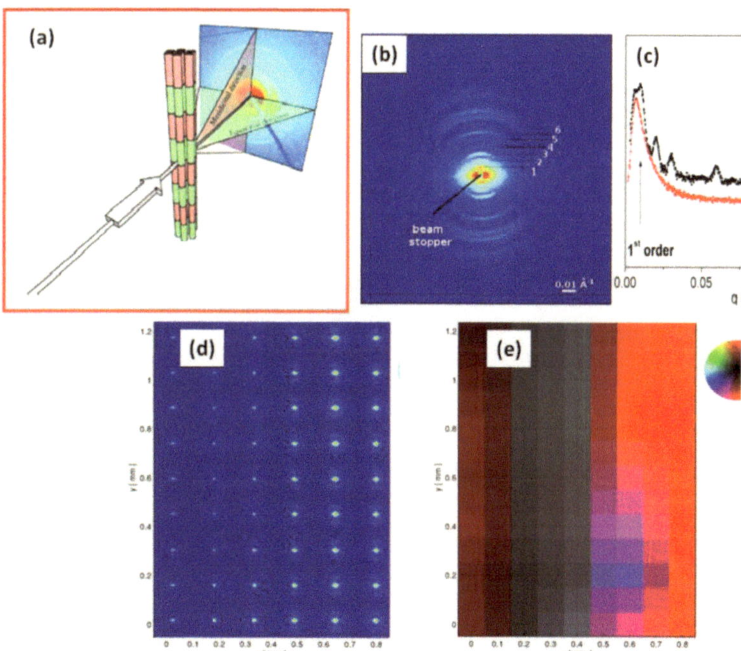

Fig. 6.15: (a) The way the fiber has to be oriented with respect to the X-ray beam, (b and c): the typical SAXS 2D image and corresponding azimuthally integrated 1D pattern; (d) a map of the previous information across a $1.2 \times 0.8\,\mathrm{mm}^2$ area, and (e): fiber orientation map converted into colors, as indicated in the color wheel. Fig. 6.15a reproduced from [36]. Fig. 6.15a–6.15b reproduced with permission of the International Union of Crystallography (http://journals.iucr.org/) from [10].

Figures 6.12–6.14 (panels b–d) and Figure 6.15(e) clearly show how the scanning microscopy with WAXS and SAXS contrast can be used to return a structural map of the actual sample, providing detailed information in an intuitive micrographic representation. These techniques allow to inspect crystalline (WAXS) and noncrystalline (SAXS) matter with atomic or nanometric resolution, respectively. This advantage was recently fully exploited with a novel type of scanning X-ray microscopies. Alternative to microscopy, in-situ SAXS and WAXS scanning experiments [36–38] were recently performed to study real time evolution processes. Accardo *et al.* [36] followed the microstructural evolution of lysozyme solution droplets on surfaces *in situ* revealing the increasing protein concentration and the onset of precipitation. Roth *et al.* [37] monitored the nanostructuring and cluster formation of gold nanoparticles deposited in aqueous solution on top of a pre-structured polystyrene colloidal thin film.
At synchrotron radiation sources, due to the increase of the available photon flux and detector performances, it can be planned to perform experiments with SAXS and WAXS scanning microscopies but also to follow in time evolving processes. For example, Campi *et al.* [38] investigated different mineralizations of a tissue-engineered bone by both spatially and temporally monitoring the bone growth and mineralization processes.

Acknowledgments

The authors are grateful for the SEED project "X-ray synchrotron class rotating anode microsource for the structural micro imaging of nanomaterials and engineered biotissues (XMI-L@B)" – IIT Protocol no. 21537 of 23/12/2009and the FIRB 2009/2010 project "Rete integrata per la Nano Medicina (RINAME)" – RBAP114AMK_006. Rocco Lassandro and Giuseppe Chita are acknowledged for the technical support.

Bibliography

[1] Cranford S, and Buehler MJ. Materiomics: Biological protein materials, from nano to macro. Nanotechnology, Science and Applications 2010; 3:127–48.
[2] Dorfs D, Krahne R, Giannini C, Falqui A, Zanchet D and Manna L. Quantum dots: Synthesis and characterization, in: Comprehensive Nanoscience and Technology (vol. 1), pp. 219–270, Elsevier B.V., Amsterdam, 2011.
[3] Jacobsen, C, Williams, S, Anderson, E, Browne, MT, Buckley, CJ, Kern, D, Kirz, J, Rivers, M, Zhang X. Diffraction-limited imaging in a scanning transmission x-ray microscope. Opt. Comm. 1991; 86: 351–64
[4] Snigireva I. and Snigirev A. X-Ray microanalytical techniques based on synchrotron radiation. J. Environ. Monit., 2006, 8, 33–42
[5] Snigirev A, Kohn V, Snigireva I, Souvorov A, Lengeler, B. Focusing high-energy X-rays by compound refractive lenses. Applied Optics 1998; 37: 653–62.
[6] Kirkpatrick P, Baez AV. Formation of optical images by X-rays. Journal of the Optical Society of America 1948; 38: 766–73.

[7] Baez AV. Fresnel zone plate for optical image formation using extreme ultraviolet and soft x radiation. Journal of the Optical Society of America 1961; 51: 405–12.
[8] Di Fabrizio E, Romanato F, Gentili M, Cabrini S, Kaulich B, Susini J, Barrett R. High-efficiency multilevel zone plates for keV X-rays. Nature 1999; 401: 895–98.
[9] Giannini C, Diaz A, Favre-Nicolin V. Coherent diffraction imaging: Basic concepts and applications, in: Diffraction at the Nanoscale – Nanocrystals, Defective & Amorphous Materials (2010), pp. 191–214, Insubria University Press, Como, 2010.
[10] Altamura D, Lassandro R, Vittoria FA, De Caro L, Siliqi D, Ladisa M, Giannini C. X-ray microimaging laboratory (XMI-LAB). J. Appl. Cryst. 2012; 45: 869–73.
[11] Als-Nielsen J, McMorrow D. Elements of Modern Physics, Wiley Ed., West Sussex, England, 2011.
[12] Cervellino A, Giannini C, Guagliardi A. Determination of nanoparticle structure type, size and strain distribution from X-ray data for monatomic f.c.c.-derived non-crystallographic nanoclusters. J. Appl. Cryst 2003; 36: 1148–58.
[13] Cervellino A, Giannini C, Guagliardi A, Ladisa M. Nanoparticle size distribution estimation by a full-pattern powder diffraction analysis. Phys. Rev. B 2005; 72: 035412.
[14] Cervellino A, Giannini C, Guagliardi A. DEBUSSY: a Debye user system for nanocrystalline materials. J. Appl. Cryst. 2010; 43: 1543–47.
[15] FOXS Fast SAXS Profile Computation with Debye Formula (http://bl1231.als.lbl.gov/2010/02/fast_saxs_profile_computation.php).
[16] Klabunde KJ (ed.). Nanoscale Materials in Chemistry, John Wiley & Sons, Inc., New York, 2001.
[17] Klabunde KJ, Stark JV, Koper OB, Mohs C, Park DG, Decker S, Jiang Y, Lagadic I, Zhang D. Nanocrystals as stoichiometric reagents with unique surface chemistry. J. Phys. Chem. 1996; 100: 12142.
[18] McCusker LB, von Dreele RB, Cox DE, Louër D, Scardi P. Rietveld refinement guidelines. J. Appl. Cryst. 1999; 32: 36–50.
[19] Full Prof Suite (http://www.ill.eu/sites/fullprof/index.html).
[20] Schaaff TG, Shafigullin MN, Khoury JT, Vezmar I, Whetten RL, Cullen WG, First PN, Gutierrez-Wing C, Ascensio J, Jose-Yacaman MJ. Isolation of smaller nanocrystal Au molecules: Robust quantum effects in optical spectra. J. Phys. Chem. B 1997; 101: 7885–91.
[21] Craievich AF. Small-angle X-ray scattering by nanostructured materials, in: Sakka S (ed.), Characterization of Sol-Gel Materials and Products (vol. 2), pp. 161–189, Kluwer Academic Publishers, Massachusetts, USA, 2005.
[22] Werner S, Wagner HD. The material bone: Structure-mechanical function relations. Annu. Rev. Mater. Res. 1998; 28: 271–98.
[23] Rho JY, Kuhn-Spearing L, Zioupos P. Mechanical properties and the hierarchical structure of bone. Med. Eng. & Phys. 2002; 20: 92–102.
[24] Olszta MJ, Cheng X, Jee SS, Kumar R, Kim YY, Kaufman MJ, et al. Bone structure and formation: A new perspective. Mat. Sci. & Eng. 2007; R58: 77–116.
[25] Dunlop JDW, Fratzl P. Biological Composites. Annu. Rev. Mater. Res. 2010; 40: 1–24.
[26] Giannini C, Siliqi D, Bunk O, Beraudi A, Ladisa M, Altamura D, Stea S, Baruffaldi F. Correlative light and scanning X-ray scattering microscopy of healthy and pathologic human bone sections. Sci. Rep. 2012; 2: 435.
[27] Hamed E, Jasiuk I, Yoo A, Lee Y, Liszka T. Multi-scale modelling of elastic moduli of trabecular bone. Journal of the Royal Society Interface 2012; 9: 1654–73.
[28] Prockop DJ, Kivirikko KI. Collagens: Molecular biology, diseases, and potentials for therapy. Annu. Rev. Biochem. 1995; 64: 403–34.
[29] Silver FH, Freeman JW, Seehra GP. Collagen self-assembly and the development of tendon mechanical properties. J. Biomech. 2003; 36: 1529–53.

[30] Ottani V, Martini D, Franchi M, Ruggeri A, Raspanti M. Hierarchical structures in fibrillar collagens. Micron 2002; 33: 587–96.
[31] Lee CH, Singla A, Lee Y. Biomedical applications of collagen. Int. J. Pharm. 2001; 221: 1–22.
[32] Ruszczak Z. Collagen as a carrier for on-site delivery of antibacterial drugs. Adv. Drug. Deliver. Rev. 2003; 55: 1595–611.
[33] Fratzl P. Cellulose and collagen: From fibres to tissues. Current Opinion in Colloid and Interface Science 2003; 8: 32–9.
[34] Gentleman E, Lay AN, Dickerson DA, Nauman EA, Livesay GA, Dee KC. Mechanical characterization of collagen fibers and scaffolds for tissue engineering. Biomaterials 2003; 24: 3805–13.
[35] Bigi, A, Ripamonti, A, Roveri, N, Compostella, L, Roncon, L and Scivazappa. Structure and orientation of collagen fibers in human mitral valve. L. Int. J. Biol. Macromol. 1982; 4: 387–92.
[36] Suhonen H, Fernandez M, Serimaa R and Suortti P. Simulation of small-angle x-ray scattering from collagen fibrils and comparison with experimental patterns. Phys. Med. Biol. 2005; 50: 5401–5416.
[37] Accardo, A, Burghammer M, Di Cola, E, Reynolds, M, Di Fabrizio, E, Riekel C. Lysozyme fibrillation induced by convective flow under quasi contact-free conditions. Langmiur 2011; 27(13): 8216–22.
[38] Roth SV, Herzog G, Körstgens V, Buffet A, Schwartzkopf M, Perlich J, Abul Kashem MM, Döhrmann R, Gehrke R, Rothkirch A, Stassig K, Wurth W, Benecke G, Li C, Fratzl P, Rawolle M, Müller-Buschbaum P. In situ observation of cluster formation during nanoparticle solution casting on a colloidal film. J. Phys: Cond. Matter 2011; 23: 254208.
[39] Campi G, Ricci A, Guagliardi A, Giannini C, Lagomarsino S, Cancedda R, Mastrogiacomo M, Cedola A. Early stage mineralization in tissue engineering mapped by high resolution X-ray microdiffraction. Acta Biomater. 2012; 8(9): 3411–8.

Roberto Comparelli, Elisabetta Fanizza, Marinella Striccoli, and
M. Lucia Curri

7 Characterization of inorganic nanostructured materials by electron microscopy

7.1 Introduction

In modern materials science the main trends are represented by fabrication, characterization and application of materials and structures with dimensions in or near the nanometer range. In this frame, inorganic nanomaterials have been increasingly attracting the attention of scientists owing to their interesting electronic, optical, magnetic, thermal, mechanical, and chemical properties, which are highly dependent on their size and shape. Such geometry-dependent characteristics are essential for their high impact on fundamental aspects as well as for the high potential in technology applications.

Properties of materials in nanophase state can strongly differ from properties of corresponding bulk phase materials also due to the substantial increase in the fraction of surface atoms and the resulting enhanced role of surface properties and interactions.

The investigation and the understanding of the properties of a wide range of materials at nanoscale have become extremely relevant, as they enable control over the material's characteristics. In particular, the synthesis and fabrication of inorganic nanomaterials have attracted increasing attention and a plethora of novel methods and procedures have been implemented and are constantly being improved to achieve an increasingly high and accurate control over size, morphology, crystalline phase and chemical nature of the resulting nanoobjects.

Among the experimental methods that have demonstrated a particular effectiveness, electron microscopy (EM) techniques represent a unique investigation opportunity.

EM offers a set of invaluable tools for the characterization of nanostructures, providing a range of different imaging modes with the ability to obtain information on elemental composition and electronic structure at the ultimate sensitivity, down to a single atom.

The intent of this chapter is to give a general illustration of the basic principle behind this class of techniques, a description of some of the most significant experimental features relevant to effectively performing a structural and morphological EM investigation and above all to provide comprehensive treatment of such extremely active and rapidly growing research fields by illustrating examples of EM investigation of nanomaterials in order to finally depict a comprehensive landscape of the potential of such techniques.

A particularly striking feature of the EM technique relies on the ability to convey a large amount of information with an exceedingly high degree of detail and in a highly efficient manner through images, namely micrographies, that ultimately comprise lots of physical and chemical information on the investigated nanomaterials and that therefore can provide exceptional insight to elucidate properties and solve problems.

7.2 Electron microscopy

Electron microscopes were developed to overcome the restrictions of optical microscopes which are limited by the wavelength of light to a resolution of 0.2 micrometers and to 500x or 1000x of magnification. In the early 1930s this theoretical limit was reached and there was a scientific need to better investigate the fine structure of matter and to obtain more detailed images of solid surfaces, requiring higher magnifications not obtainable using light microscopes.

The first prototype of the electron microscope with a resolving power greater than that of light was the transmission electron microscope built by Ernst Ruska in 1931 (Nobel Prize in 1986) who obtained images of copper and gold surfaces using an electron beam normally incident on the sample surface, with a magnification of only 10x. In the meantime, during the 1930s a very different way of imaging solid samples, scanning electron microscopy, was invented by Knoll (1935) that applied the scanning principle for the study of television camera tubes. He was successful not only in achieving higher magnification but also in eliminating the intrinsic chromatic aberration of the electron microscope. Soon afterwards, in the USA, Zworykin *et al.* (1942) developed a dedicated SEM. Their studies were followed by a Cambridge group in the 1950s and early 1960s, headed by Charles Oatley, and finally led to the marketing of the first commercial instrument by the Cambridge Scientific Instrument Company named "Stereoscan" in 1965.

The advantages of the electron microscope with respect to the optical microscope can be basically summarized (Tab. 7.1) in terms of higher magnification, larger depth of field (or focus) and greater resolution.

A clear understanding of the meaning of magnification, depth of field and resolution allow to appreciate the potential behind this technology in the field of characterization of nanostructured materials.

Magnification is the process of enlarging an object only in appearance not in physical size. While in an optical microscope magnification is a function of the power of the objective lens, in an electron microscope electromagnetic lenses are used, namely condenser and objective lenses, which focus the beam to a spot on the specimen or on the detector. In particular in SEM the magnification is a function of the ratio between the scan area and the area of the detector, the latter being fixed it thus depends only on the size of the scan area.

Tab. 7.1: Main characteristics of the optical microscope, scanning electron microscope and transmission electron microscope. The depth of focus is reported for the high magnification value.

	Magnification	Depth of Field	Resolution
Optical Microscope	4x–1400x	0.2 μm	0.2 mm
Scanning Electron Microscope	10x–1Mx	1.0 μm	1.5 nm
Transmission Electron Microscope	50x–1.5Mx	0.2 μm	0.1 nm

Depth of field (or focus) is the vertical range in which sample features remain in focus. Focusing is defined as adjusting the imaging conditions in order to get the image as sharp as possible (the image is in-focus). In the observation of a specimen with a substantial depth, if the focus is adjusted to the top side, the bottom side may be out of focus. In such a case, if the range between the upper and lower image blur is large, it is said that "the depth of focus is large", whereas if the range is small, it is said that "the depth of focus is small".

Resolution is the ability to resolve two closely spaced points. Resolution is in an inverse relation to wavelength. Indeed, the use of an electron beam instead of light as the probe dramatically improves the resolution power Actually, also other operating condition should be carefully chosen in order to achieve high resolution.

Although the electron microscope has had, since its infancy, an enormous impact in materials and life science, it was the continuous and massive development carried out throughout the following decades that has impressively affected its application to a vast variety of materials, and ultimately strongly empowered the investigation of crystalline nanomaterials. One of the typical characteristics of nanomaterials consists in the very small particle dimension. Although their size dependent properties can be indirectly studied by means of a variety of techniques, depending on the specific feature of the nanoparticle (NP) composition, including spectroscopies, X-ray and neutron diffraction, direct imaging of NPs is only possible using transmission or scanning electron microscopy (TEM and SEM, respectively).

Among the other relevant features, the ultimate resolution, for instance, has improved in the last years down to the typical inter-atomic spacing of most of the important class of materials.

7.2.1 Working principles

The basic scheme of the electron microscope is similar for TEM and SEM (Fig. 7.1) and works mimicking exactly its optical counterpart, except for the use of a focused beam of electrons instead of an optical radiation to "image" the sample surface and gain information on its structure and composition. In electron microscopes a flow of electrons is generated in high vacuum by a source of electrons (electron gun). This flow is accelerated toward the sample by means of a positive electrical potential, while the size of

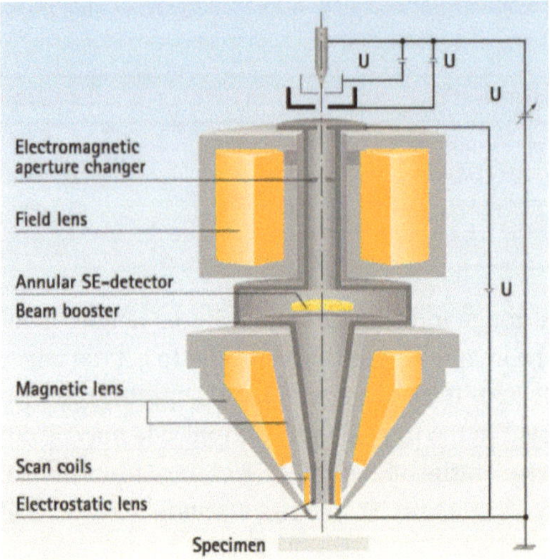

Fig. 7.1: Basic scheme of an electron microscope (courtesy of Zeiss).

the beam is controlled with metal apertures and electromagnetic lenses, to obtain a focused and monochromatic beam.

Electromagnetic lenses can be classified in **condenser lenses**, which focus the electron stream into a small and coherent beam, and **objective lenses**, which focus the beam on the specimen (in SEM) or on the detector (TEM).

Generally speaking, an **electromagnetic lens** consists of a coil of copper wires inside an iron pole piece. An electric current through the coils creates a magnetic field in the bore of the pole pieces which is used to converge the electron beam. When an electron passes through an electromagnetic lens it is subjected to two vector forces at any particular moment: a force (H_Z) parallel to the core (Z axis) of the lens; and a force (H_R) parallel to the radius of the lens. These two forces are responsible for two different actions on the electrons, spiraling and focusing, as they pass through the lens. An electron passing through the lens parallel to the Z axis will experience the force (H_Z) causing it to spiral through the lens. This spiraling causes the electron to experience (H_R) which causes the beam to be compressed toward the Z axis. The magnetic field is inhomogeneous in such a way that it is weak in the center of the gap and becomes stronger close to the border. Electrons close to the center are less strongly deflected than those passing the lens far from the axis. In the optical microscope the exchange of lenses allows to change magnification and the adjustment of the distance among the lenses along the optic axis allows to modify the focus. Electromagnetic lenses in contrast are stationary, but their strength can be varied, altering the amount of current passing through them. Thus electromagnetic lenses have variable focal points.

The **condenser lens** and **aperture** are set in order to achieve a thin and coherent beam. An increased strength of the condenser lens results in a smaller probe diameter on the specimen, which enables better resolution but it comes at a cost. The stronger the condenser lens setting, the larger the portion of the beam stopped by the objective aperture and thus a reduction in probe current occurs. An adequate current is thus essential to produce images with the proper contrast and signal to noise ratio.

Objective lens magnetic field, objective lens aperture size, electron acceleration voltage and scan rate may be accurately set in order to improve the resolution, to decrease phenomena causing artifacts (such as astigmatism, sample charging, *etc.*) and to suitably image the nanosized features of the object.

The role of the **objective lens** is to focus the scanning beam onto a defined area of the specimen, in a SEM, or to focus scattered electrons on the detector, in a TEM. The strength of objective lens can be controlled by controlling the intensity of the magnetic field. In particular, the magnetic field intensity changes by changing the current in the objective lens, and therefore the focal length of the lens can be tuned.

If the condenser and/or objective (and diffractive) lenses are not completely symmetrical, objects will be focused to different focal planes providing an **astigmatic image**, *i.e.*, a distorted image. Due to astigmatism the image cannot be totally sharp, and has a different amount of defocus in different directions. This can be prevented by having a well-fabricated lens but other defects such as dirt on an aperture *etc.* can also be a cause of astigmatism. In the optical microscope, astigmatism is corrected by adding corrective lenses into the optical path. In electron microscopes, the defect is corrected using a **stigmator** *i.e.*, a ring of electromagnets positioned around the beam to "push" and "pull" the electron stream to make it almost perfectly circular.

Once the beam impinges the sample, diverse physical interactions (elastic and inelastic scattering phenomena) occur in the area of incidence, depending on the electron acceleration voltage and on the sample composition. These interactions and effects can be detected and transformed into an image by means of a proper detection system (fluorescent screens, scintillators, CCD cameras).

7.3 Scanning electron microscopy

The development of a wide variety of materials in the nanodomains and the possibility to grow or deposit them on surfaces or assemble them in more complex structures requires powerful characterization tools. SEM plays a key role in the imaging and characterization of nanostructured materials. Its versatility is due to the improved resolution capacity with respect to the optical microscope, high depth of focus, use of different contrast mechanisms for the creation of images and adaptation and control of additional detectors for micro-range analysis.

The first and fundamental part of the SEM is the source of electrons. In a typical instrument, the electron gun consists in a V shaped filament made of tungsten or lanthanum hexaboride (LaB_6) enclosed in an electrode (Wehnelt Cap). Tungsten is commonly used thanks to its high melting point, low vapor pressure, and low cost, while LaB_6 requires a higher vacuum rate. Due to the negative potential of the electrode, a small area of the filament (point source) emits monochromatic electrons by a thermo-ionic effect. Two different types of electron guns are usually used: the standard electron guns and the field emission guns (FEG).

In standard electron guns (Fig. 7.2), the filament (cathode) is heated until a stream of electrons is produced, while a positive electrical potential is applied to the anode. The effect of the electrical potential is to accelerate the electrons toward the optical axis of the column and collect them in an area, called space charge, between the tip and the cap. Such a cap is provided with a small aperture that allows the transit of the electron beam in the column.

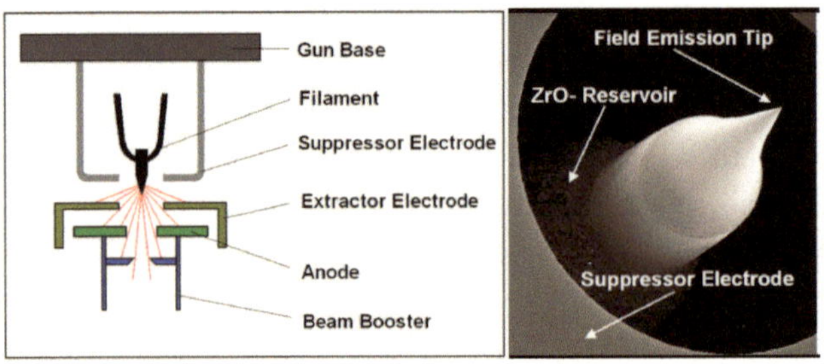

Fig. 7.2: Basic scheme of an electron gun and image of field emission gun (courtesy of Zeiss).

A field emission gun (FEG) consists of a sharp tungsten tip (about 100 nm), held at a negative potential of several kV with respect to the close anode. The difference in electrical potential is sufficient to induce the electron emission from the tip by the tunnel effect. The emitters can be made of a single tungsten crystal (cold cathode) or of a tungsten tip coated with zirconium oxide (Schottky emitter) in which the thermo-ionic effect is associated to the high electric field to produce an intense electron beam. A FEG allows to obtain sources characterized by higher brightness, monochromaticity and spatial resolution than in conventional guns (electron current >1000 times) and high source lifetime, while an ultrahigh vacuum level ($\sim 10^{-7}$ Pa) is required.

The generated electron beam, with energy ranging from 0.2 keV to 40 keV, is then focused by a series of electromagnetic lenses in conjunction with two apertures to a spot about 0.2 nm to 5 nm in diameter. A set of scan coils sweep the beam across the sample by using a magnetic field generated by a time-varying voltage applied to the coils. The sweeping occurs in a grid pattern called **raster** over a rectangular area of the sample surface, similar to the raster in a television receiver. The same varying voltage sent to the coils is also applied to the deflection scanner of a display.

The electron beam, incident on the sample, generates a complex series of interactions and induces the emission of electrons and photons from the sample surface. Several detectors can be used to reveal the electron emissions, producing a voltage signal which is amplified and then transferred to the display unit. Thanks to the synchronization between the display and the electron probe scan, brightness variations, depending on the number of emitted electrons from the sample surface, appear on the display and form the SEM image, following the topography of the sample. The image can be recorded in digital format and properly processed.

7.3.1 Magnification and resolution of SEM

One of the main advantages of the EM with respect to the optical microscope is the possibility to have highly improved resolution and magnification. The resolution is defined as "the minimum distance that can be separated as two distinguishable points in the image". In a SEM, the resolution is determined by the minimum spot size of the electron beam incident on the specimen surface, which, in turn, depends on both the wavelength of the electrons and the electron-optical system that produces the scanning beam. The resolution is also limited by the size of the interaction volume, or the extent to which the material interacts with the electron beam. Values as low as 1 nm are possible in a SEM, depending on the electron gun, acceleration voltage and sample surface.

While in an optical microscope and in a TEM the magnification ratio is strongly dependent on the objective lens, in a SEM, the linear magnification is expressed by the ratio between the scan length of the display unit monitor and the scan size on the sample surface. Then the magnification can be improved, reducing the scan size on the sample or increasing the monitor diameter. Magnification can be therefore controlled by tuning current supplied to the x, y scanning coils, over a range of up to 6 orders of magnitude from about 10 to 1000 times.

7.3.2 Interaction of the electron beam with the sample: elastic and inelastic scattering

When an energetic and focused electron beam interacts with the specimen surface, the incident electrons can undergo elastic or inelastic scattering processes. Indeed, the incoming electron can interact with one of the nucleus of an atom forming the sample and be backscattered with virtually undiminished energy, or it can interact with the orbiting electrons of sample atoms in a variety of ways, gradually losing its energy in the process. These interactions are not limited to the surface layers but, depending on the electron beam energy, the atomic number of the specimen material, and its density, the penetration depth can reach a few micrometer under the surface, defining a "tear drop" shaped interaction volume. Each type of interaction potentially constitutes a signal that carries information about the sample. For instance, the most frequent interaction is the ejection of an electron from the atom with relatively low energy, less than 50 eV. If this occurs near the sample surface, the liberated electron may escape and be detected as a secondary electron (SE). Also backscattered electrons (BSE) can cause the emission of secondary electrons as they travel through the sample and exit the sample surface. Other signals include characteristic X-rays, cathode-luminescence, and absorbed current, each carrying a specific type of information (Fig. 7.3).

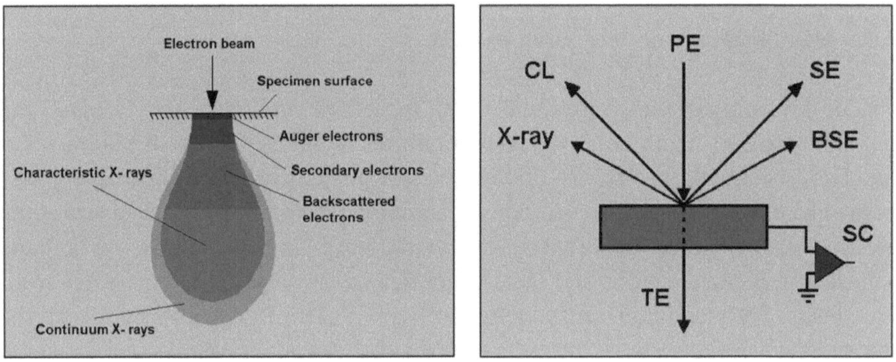

Fig. 7.3: Primary interactions between the incident electron beam and the sample surface in SEM (courtesy of Zeiss).

7.3.2.1 Secondary electrons and their detection

Secondary electrons (SE) are produced when an incident electron excites an electron in the sample and loses most of its energy in the process. The excited electron moves inside the sample undergoing elastic and inelastic collisions until it reaches the surface, from where it can escape if it still has sufficient energy, being then scattered in a large angle. The emission yield of SEs depends on many factors, and is generally higher for high atomic number materials, and at higher angles of incidence. This effect is important for image resolution and contrast. The energy loss from inelastic scattering is inversely related with the energy, then SEs originate within a few nanometers from the sample surface. Therefore, the production of SE is closely related to the topography. Due to their low energy (5 eV) only SEs that are very close to the surface (< 10 nm) can exit the sample and be detected. Any changes in topography of the sample occurring in a scale larger than this sampling depth will change the emission yield of SEs. Collection of these electrons is aided by using a "collector" in conjunction with a secondary electron detector. An SE detector is an Everhart–Thornley type of detector, consisting of a CaF scintillator with a surrounding Faraday cage used to both attract SEs as well as to shield the electrons from the high voltage applied to the scintillator itself. The SEs are first collected by attracting them towards an electrically biased grid at about + 400 V, and then further accelerated towards the scintillator, positively biased to about + 2,000 V. The accelerated SEs are then sufficiently energetic to cause the scintillator to emit flashes of light which are conducted to a photomultiplier outside the SEM column *via* a light pipe and a window in the wall of the specimen chamber. The brightness of the signal depends on the number of secondary electrons reaching the detector. For normal incidence of the electron beam, the interaction area is symmetric and the SEs will be emitted isotropically around the axis. When the angle of incidence increases, *e.g.*, for samples with edges and steps, the average path of the SEs in the specimen decreases, inducing more SEs to be emitted. Thus morphological irregularities tend to look brighter than flat surfaces, resulting in images with a three-dimensional appearance. Moreover, tilting the specimen facing the detector is advantageous as it improves the SE detection efficiency.

In order to increase the detection efficiency, a new kind of detector has recently been developed and is used in modern equipment. The electrons generated in the upper range of the interaction area that, therefore, contain direct information on the sample surface, can be detected very efficiently by using the **in-lens detector**, whose detection efficiency results from its geometric position in the beam path and from the combination with the electrostatic/electromagnetic lens. Indeed, the detector is placed above the objective lens (Fig. 7.4). In this way the electrostatic field of the lens induces an acceleration to the SEs generated on the sample surface that are focused through the electromagnetic field to the in-lens detector. The efficiency of the in-lens detector is then largely determined by the electric field of the electrostatic lens. As the intensity of the electrostatic field decreases exponentially with the distance, the working distance (WD, which is defined as the distance between the final objective lens and the sample surface) is one of the most important factors to determine the signal/noise ratio and thus the efficiency of such a detector.

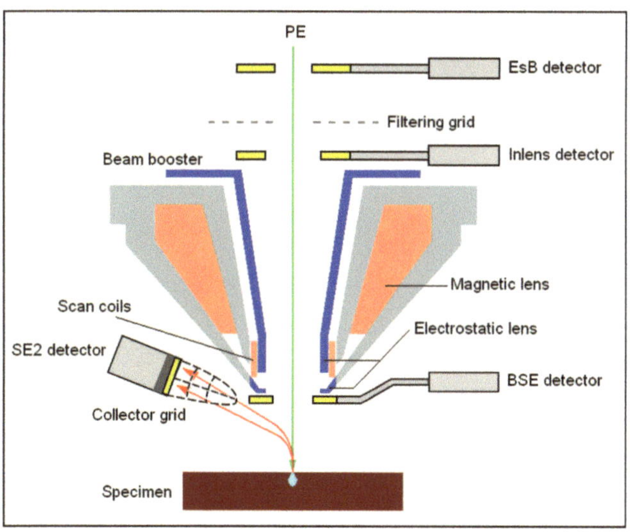

Fig. 7.4: Scheme of the objective lens system, illustrating the geometrical positioning of conventional off-lens SE and in-lens detectors (courtesy of Zeiss).

7.3.2.2 Backscattered electrons and their detection

If no loss of energy is involved, the scattering of the electron beam is considered an elastic Rutherford scattering. Backscattered electrons (BSE) consist of high energy electrons originating in the electron beam that are reflected or backscattered out of the specimen interaction volume. The production of BSE varies according with the atomic number of the element forming the specimen, and it is higher for higher atomic numbers. Such different production rates cause higher atomic number elements to appear brighter than lower atomic number elements. This interaction can be used to make a distinction among parts of the specimen formed of atoms with a different average atomic number. Therefore, BSE can be used to detect contrast between areas with different chemical compositions. However, the distribution profile of the BSE is peaked in the forward direction and decreases dramatically with the increase of the angle of scattering. For this reason, Everhart–Thornley detectors, normally positioned to one side of the sample, are partially effective for the detection of BSE because only few such electrons would be collected in the solid angle subtended by the detector, and because the positively biased collector has a low probability to attract them. Indeed, although it is still possible to set the voltage of the Everhart–Thornley detectors to a negative value which will improve the yield of BSE detected, a specific BSE detector is recommended if imaging of the material contrast is required. BSE detectors are positioned above the sample, coaxially with the electron beam which allows to increase the electron collection rate. For the geometrical reasons reported above, a hole is present in the center of the detector to allow the scan of the electron beam (Fig. 7.5). BSE detectors are usually scintillators or semiconductor devices such as solid state diodes. Detector efficiency is mainly determined by the acceleration voltage, the selected WD and the orientation of the sample towards the detector. Indeed, its efficiency increases with increasing acceleration voltage and decreases when the WD is too low or too high, as only few BSE will

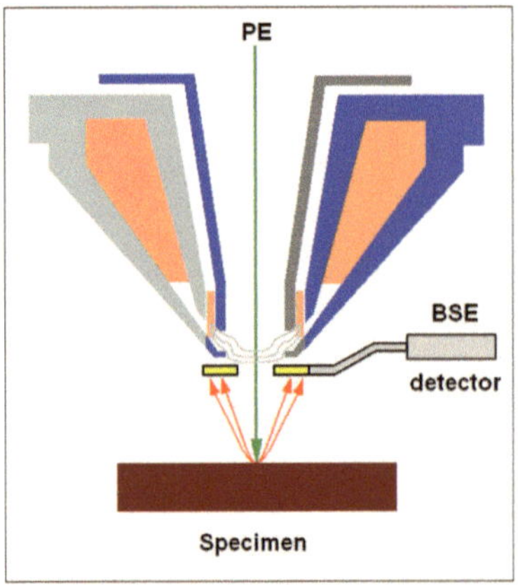

Fig. 7.5: Scheme of the objective lens system, illustrating the geometrical positioning of BSE detectors (courtesy of Zeiss).

then hit the detector. The orientation of the specimen facing the detector is decisive for high efficiency BSE collection.

7.3.2.3 Energy loss

The electron beam can interact with the sample, transferring energy to the electrons of the inner shells of the specimen atoms, resulting in ionization processes or electronic rearrangement. The atom that absorbs the energy can either emit a photon (radiative process) or eject an electron from the same or from a different shell (nonradiative Auger process). The radiative transitions are generally in the X-ray region of electromagnetic spectrum and lead to emission of photons mainly from K, L, M-shells. These X-rays, which are characteristic of the elements and the X-ray spectra, can provide a fingerprint of the composition of the specimen and of the relative abundance of elements in the sample. The energy dispersive X-ray spectrometer (EDS also called EDX) measures the number of emitted X-rays versus their energy. To get an absolute value (number of atoms of one type as a fraction of all atoms) normalization of the emission yields has to be performed by using a standard sample under identical conditions so that ratios can be formed. The X-ray detector is a lithium-drifted silicon [Si(Li)] crystal detector. The detector is kept at low temperature by using liquid nitrogen or a Peltier cooling system in order to reduce leakage current and to prevent possible electrical breakdown in the crystal, induced by the applied bias voltage due to decreased resistance at higher temperatures

Also photons in the visible range of the electromagnetic spectrum can be emitted upon interaction of the electron beam with the sample atom. Cathode-luminescence (CL) is the emission of photons of characteristic wavelengths from a material that is under high energy electron bombardment. Concerning the nonradiative processes, Auger electrons are characteristic of the fine structure of the atom and have energies between 280 eV (carbon) and 2.1 keV (sulfur). By discriminating between Auger electrons of various energies, a chemical analysis of the specimen surface can be made by using Auger electron spectroscopy tools (AES).

7.4 Transmission electron microscopy

TEM is a unique tool as it provides a real space image of the atom distribution in the nanocrystal (NC) and on its surface exploiting electrons transmitted by the sample. In addition, currently TEM, with its more advanced capabilities, offers not only atomic resolution lattice images, but also chemical information at 1–2 nm allowing direct identification of the chemistry of a single NP. Structural characteristics of a single NP can be fully explored by a fine tuning of the electron probe. Scanning transmission microscopes (STEMs) are distinguished from conventional transmission electron microscopes in that the narrow spot arising from the focused electron beam on the specimen is scanned over the sample in a raster.

7.4.1 The instrument

A modern TEM can be schematically described as composed of an illumination system, a specimen stage, an objective lens system, a magnification system, data recording systems and chemical analysis systems. The illumination system is based on generation of an electron beam. There are two different ways to produce electron beams: by using electron guns, thus achieving electronic emission through heating, hence a thermo-ionic emission, typically by using W or – most commonly – LaB6, or, alternatively, by using a field emission source in which electron emission occurs through applying an extraction voltage (see Section 7.3). The thermo-ionic gun provides a high illumination current but the current density and the beam coherence are not as high as those of a field emission gun. FEG is unique for performing high coherence lattice imaging, electron holography and high spatial resolution microanalysis.

Once the electrons are generated, the stream needs to be accelerated to the required high energy and this is performed by applying a positive potential to an **extractor anode** as in SEM.

The illumination system also includes the **condenser lenses** that are crucial to generate a fine electron probe. A set-up of different magnetic lenses and apertures makes it possible to get either a parallel beam (micro-probe for TEM) or a convergent beam with selected convergence angles (nanoprobe for scanning transmission electron microscopy, STEM). Furthermore, the beam can be not only scanned as in STEM, but also tilted as in dark field TEM (DF TEM).

Specimen stage is a key for carrying out structure analysis because it can be used to perform in-situ observation of phenomena induced by annealing, electric field or mechanical stress, giving the possibility to characterize the physical properties of individual nanostructures.

The **objective lens** is the most important lens in the microscope since it generates the first intermediate image whose quality determines the resolution of the final image. Conversely to SEM, the objective lens is placed below the sample. The magnification system consist of intermediate lenses, that can be switched with diffraction lens to switch between imaging and diffraction mode, and projection lenses, that provide a further magnification of the second intermediate image (image or diffraction pattern, respectively). The whole system can reach a magnification up to 1.5 million.

Images and diffraction patterns can be directly observed on the viewing screen in the projection chamber or *via* a TV camera mounted below the microscope column. Images can be recorded by means of different systems, such as on negative films, on slow-scan CCD cameras or on imaging plates.

Because of strong interactions of electron with matter, gas particles must be absent in the column, therefore a high vacuum is required and is maintained by a vacuum system typically comprising a rotary pump (pre-vacuum pump), a diffusion pump and one or more ion pumps.

7.4.2 Image formation process

In order to understand the image capability of the technique, and effectively take advantage of its full potential, it is worthwhile illustrating the image formation process in a TEM.

According to Abbe's theory of image formation, the microscope image is the interference effect of a diffraction phenomenon, namely image formation depends on interference between nondiffracted light and diffracted light. In this sense, a lens serves to recombine light diffracted from the specimen such that the diffracted light interferes with itself in a manner that recreates an image of the specimen.

The resolution of a TEM is mainly determined by the objective lens. The surface of a specimen is illuminated by a parallel (or nearly parallel) electron beam. The electron beam is diffracted by the lattice of the crystal, forming the diffracted beams which can propagate along different directions. The electron-specimen interaction can be explained by taking into account distinct phenomena (Fig. 7.6).

When elastic scattering of the electrons of the beam occurs, due to interaction with the atoms of the specimen, the electrons penetrating into the electron cloud of an atom of the sample can be thought of as attracted by the positive potential of the nucleus (Coulomb interaction), and their path is changed as a result.

The closer the electron comes to the nucleus, the higher the attractive force and consequently the scattering angle. In rare cases, even complete backscattering can occur (backscattered electrons BSE). These interactions can be treated as elastic, which means that no energy is transferred from the scattered electron to the atom. The mass-

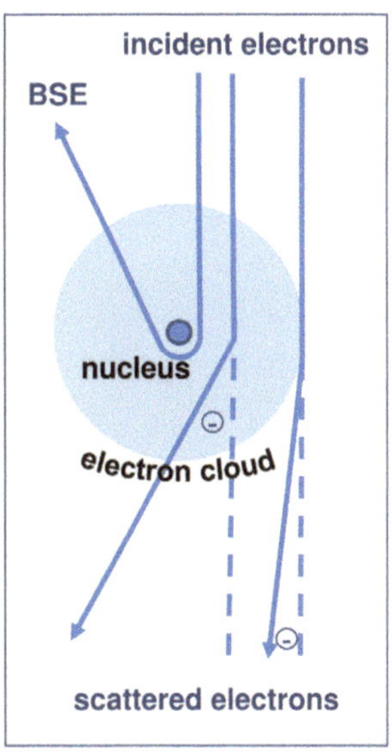

Fig. 7.6: Scheme of elastic scattering of electrons by atoms.

thickness contrast is somehow related to the contrast in optical microscopy, but it is the local scattering power that determines contrast in TEM instead of absorption of light. The interaction of electrons with heavy atoms is stronger than with light atoms so that areas in which heavy atoms are localized appear with darker contrast than such with light atoms (mass contrast). In thick areas, more electrons are scattered of course; these areas appear dark (thickness contrast). In particular, the mass-thickness contrast is important in bright and DF imaging. The strong Coulomb interaction of the electrons with the potential atom core, which leads to high angle scattering (designated as Rutherford scattering), is employed by Z-contrast imaging in STEM. By this method, small clusters (or even single atoms) of heavy atoms can be imaged in a matrix of light atoms since the contrast is proportional to Z^2 (Z: atomic number). If a crystalline specimen is transmitted by electrons, then Bragg diffraction happens as well. Each atom in such a regular arrangement acts as a scattering center. The scattered electron waves may interact with each other in a constructive or destructive way, which gives rise to a diffraction pattern. If a crystal is oriented along a zone axis so that many electrons are strongly scattered to contribute to the reflections in the diffraction pattern, then only a few electrons pass without interactions and therefore this crystal appears with dark contrast in the bright field (BF) image (diffraction contrast). In real specimen, all contrast mechanisms, namely mass-thickness and Bragg contrast, occur simultaneously, making interpretation sometimes difficult.

Electrons that come from the condenser system of the TEM are scattered by the sample situated in the object plane of the objective lens (Fig. 7.7).

Electrons scattered in the same direction are focused in the back focal plane, and, as a result, a diffraction pattern is formed. Electrons coming from the same point of

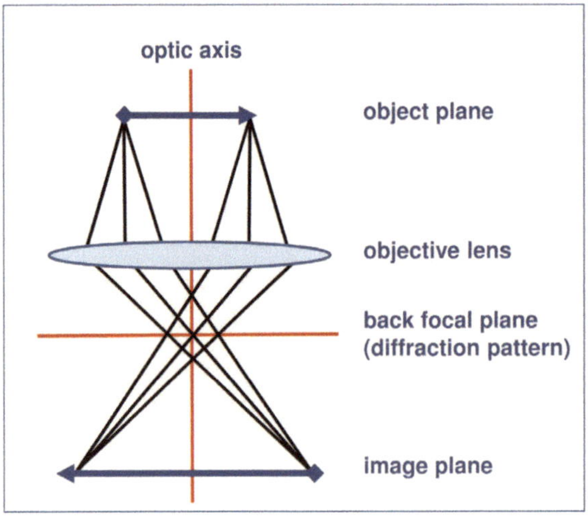

Fig. 7.7: Representation of the ray path leading to the first intermediate image in the image plane of the objective lens.

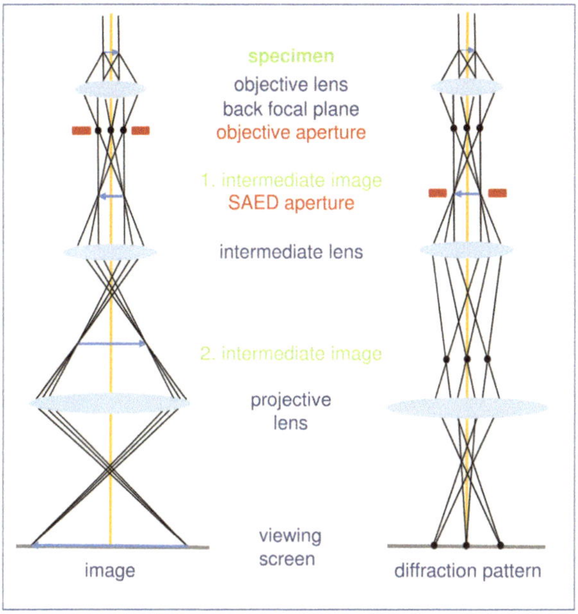

Fig. 7.8: Abbe theory for the image formation in a TEM (left) scheme for diffraction pattern.

the object are focused in the image plane (Fig. 7.8). In the TEM, the first intermediate image is magnified by further lenses (**projection lens**).

In the **BF mode** of the TEM, an aperture is placed in the back focal plane of the objective lens which allows only the direct beam to pass (Fig. 7.9). In this case, mass-thickness and diffraction contrast contribute to image formation: thick areas, areas in which heavy atoms are enriched, and crystalline areas appear with dark contrast. In spite of the useful information obtainable from BF images, it should be mentioned that the interpretation of contrast is often impeded since these phenomena occur simultaneously.

In **DF images**, one or more diffracted beams are allowed to pass the objective aperture. The direct beam is blocked by the aperture (Fig. 7.10). In contrast to the direct beam, the diffracted beam has interacted strongly with the specimen, and often very useful information is present in DF images, *e.g.*, about planar defects, stacking faults or particle size.

An electron microscope working in STEM mode can be also equipped with three different types of detectors (Fig. 7.11):

- the **BF detector** that is placed at the position of the aperture in BF TEM and detects the intensity in the direct beam from a point on the specimen;
- the **annular dark field (ADF)** detector is a disk with a hole in its center where the BF detector is installed. The ADF detector uses scattered electrons for image formation, similar to the DF mode in TEM. The measured contrast mainly results

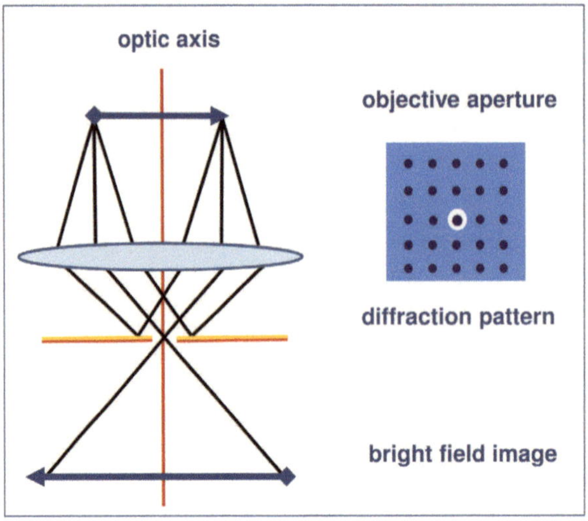

Fig. 7.9: Scheme of bright field image formation.

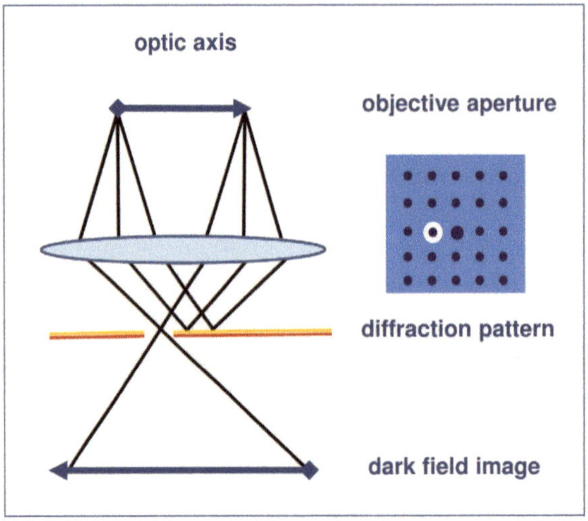

Fig. 7.10: Scheme of dark field image formation.

from electrons diffracted in crystalline areas but is superimposed by an incoherent Rutherford scattering;

- the **high-angle annular dark field** detector **(HAADF detector)** is also a disk with a hole, but the disk diameter and the hole are much larger than in the ADF detector. Thus, it detects electrons that are scattered to higher angles and only incoherent Rutherford scattering contributes to the image. Therefore, pure Z contrast is obtained.

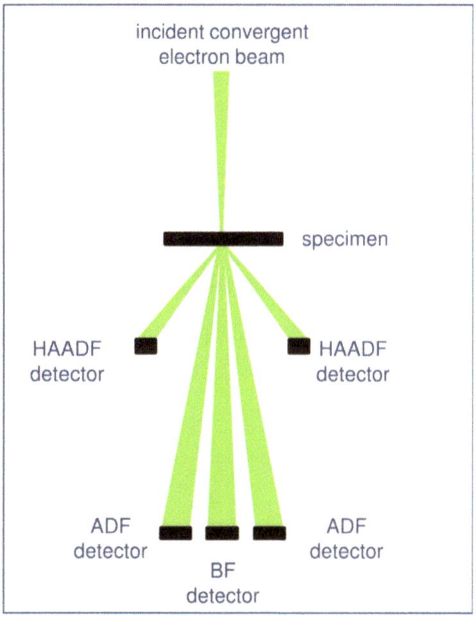

Fig. 7.11: Arrangement of the various STEM detectors (schematic outline).

Electron diffraction (ED) is a collective elastic scattering phenomenon with electrons being scattered by atoms in a regular array (crystal). The incoming plane electron wave interact with the atoms, and secondary waves are generated which interfere with each other. This occurs either constructively (reinforcement at certain scattering angles generating diffracted beams) or destructively (extinguishing of beams). As in X-ray diffraction (XRD), the scattering event can be described as a reflection of the beams at planes of atoms (lattice planes). The Bragg law gives the relation between interplanar distance d and diffraction angle

$$\Theta : n\lambda = 2d\sin\Theta.$$

Since the wavelength λ of the electrons is known, interplanar distances can be calculated from ED patterns. Furthermore, information about crystal symmetry can be obtained. Consequently, electron diffraction represents a valuable tool in crystallography.

The estimate of scattering angles is $\lambda_{el} = 0.00197\,\text{nm}$ (1.97 pm) for 300 kV electrons. A typical value for the interplanar distance is $d = 0.2\text{nm}$. If these values are put in the Bragg law, then the scattering angle is: $\Theta = 0.28°$. As a rule, the scattering angles in ED are very small: $0 > \Theta > 1$.

From this it follows that (i) the reflecting lattice planes are almost parallel to the direct beam (Fig. 7.12, left figure), (ii) the incident electron beam is the zone axis of the reflecting sets of lattice planes (Fig. 7.12, right figure).

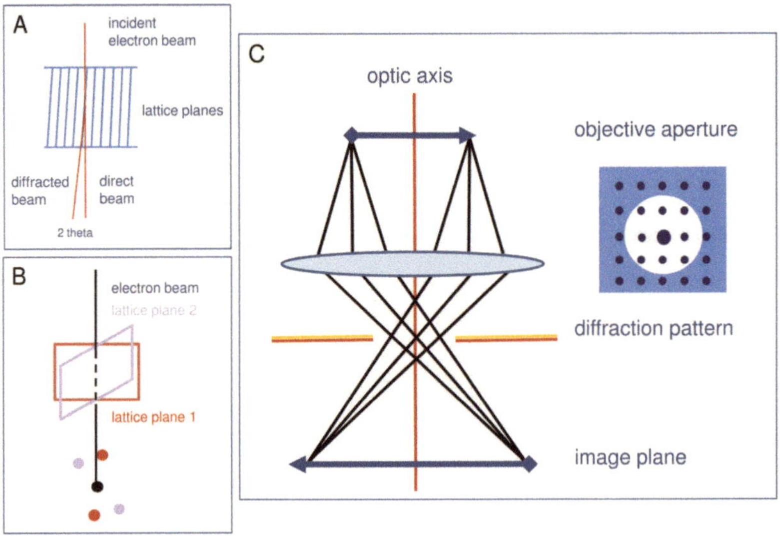

Fig. 7.12: (a) The reflecting lattice planes are almost parallel to the direct beam, (b) the incident electron beam is the zone axis of the reflecting sets of lattice planes, (c) scheme of diffraction pattern formation.

7.5 Sample preparation for electron microscopy

An effective EM investigation, performed by any technique, strongly depends on the quality of the specimens examined. For this reason, a wealth of scientific literature that has carefully detailed specific preparation steps and numerous excellent books on the preparation of thin specimens are present in the literature.

The selection of the *right sample preparation technique is the "conditio sine qua non" for its correct microscopic observation.* In this section a brief overview of EM sample preparation techniques will be provided, paying particularly attention to those useful for inorganic NC investigation. More specific publications on the subject are suggested for a more detailed discussion [1, 2].

7.5.1 SEM sample preparation

The preparative procedures for SEM investigations of nanostructured materials are strictly dependent on the electrical conductive properties of the specimens.

7.5.1.1 Casting

When the sample is a conductor, such as metal or semiconductors, no specific preparation is typically required for surface topography investigations. The easiest way to prepare an inorganic NC sample for a SEM investigation is suspending the powder in volatile solvent such as ethanol, methanol or less polar solvent such as chloroform, hexane or toluene according to the solubility of the sample, casting a few drop of the solution/suspension onto a conductive slide (*i.e.*, silicon, ITO, gold) and letting the solvent dry. The slides are then mounted onto stainless-steel sample holders by double-sided carbon tape and grounded by silver paste to avoid charging effects.

7.5.1.2 Ion sputtering

When the sample is an **insulator** and it is difficult to obtain suitable resolution even using a low acceleration voltage, then the general approach consists in coating the specimen surface with a thin layer (20–30nm) of metal. The effect of such a coating procedure is multifold. Indeed, the metal coating (i) increases conductivity of the sample, thus minimizing sample charge-up, which results in deflection of the incident beam and severe degradation of the final image, (ii) increases mechanical stability of the sample due to a better heat conduction, (iii) increases the primary and secondary electron emission, (iv) decreases the beam penetration, resulting in a higher spatial resolution.

Metal deposition on a sample surface is commonly performed by **ion sputtering**.

The working principle of ion sputtering is based on the bombardment of a target with fast heavy particles that induce the erosion of the target material. The process occurs in conditions of gaseous glow discharge between an anode and a cathode. The resulting omni-directional deposition of sputtered atoms will form an even coating on the surface of the specimen. Enhancement of this process for SEM sample coating is obtained by the choice of a suitable ionization gas (generally argon or nitrogen) and target material.

A common target material is gold. However, for FEG-SEM applications, where an extremely thin, grain-free, uniform coating is needed to eliminate charging and to improve contrast on low density materials, different metals and alloys can also be used, such as Au/Pd for high resolution coating, Cr for semi-conductor materials and high resolution backscattered electron imaging, and Ir that allows to obtain a virtually grain-free coating material.

While metal coatings are useful for increasing signal to noise ratio, they are not desirable when X-ray microanalysis is employed. For this reason, when using X-ray spectroscopy a carbon coating is preferred, unless the element under consideration is carbon. In that case aluminum can be used.

7.5.2 Sample preparation for TEM

The golden rule to select the best preparation technique for a sample is choosing the one that produces a suitable thin slice of the material to be investigated. The technique must also be suitable for the different TEM analyses and should contain a minimum of artifacts. Due to the strong interaction between electrons and matter, the specimens have to be rather thin, from 40 to 150 nm. This is achieved with several methods, depending on the material.

Among the many preparation techniques for thin slices, a distinction should be made between **direct** and **indirect** preparation methods. **Direct methods** are used to analyze thinned materials or fine particles on a TEM support grid in order to perform structural, chemical, or spectroscopic analysis. **Indirect methods** do not enable direct observation of the material, as they involve the observation of an imprint of the material made using carbon films deposited on its surface, either with or without the shadowing effect of metallic particles (some examples are direct and indirect-replica techniques and freeze-fracture techniques) [2].

Here the focus will be made on direct methods **specific to fine particle analyses** such as inorganic NCs. Such techniques include, among others which will not be reported here, **fine particle dispersion** and **ultramicrotomy**.

The **fine particle dispersion technique** is used to prepare a suspension of any NC in order to obtain a slide or a film of thickness enabling TEM observation. These include **casting** and **ultramicrotomy**. The easiest way to observe NPs is depositing them on a TEM support film by simply **casting** a few drops of the suspension onto thin holey carbon support film and letting the solvent evaporate. No long and involved grinding, polishing, or ion milling is required, making TEM a rapid means for nanostructure characterization. Such an approach results in random orientations of the NPs on the TEM grid.

For **ultramicrotomy**, the orientation can either be random or not depending on the shape of the particles (spheres, platelets, fibers, *etc.*) and on the possibility to orient them. In this case, NPs must be embedded in a resin in order to make a block that can be handled and then sectioned using the ultramicrotomy technique. Ultramicrotomy can be also carried out under **cryo** conditions *i.e.*, cooling the material under liquid nitrogen, for instance. **Cryo-ultramicrotomy** is mainly used in the investigation of biological materials, or for the preparation of thin slices of polymers.

Ultramicrotomy techniques could be useful for instance to observe **nanocomposite** materials, such as multiphase materials containing at least one nanocrystalline phases. A typical example of nanocomposite is obtained by embedding NCs in organic matrixes as polymers. Mixed polymer–nanomaterials samples can be prepared by ultramicrotomy or cryo-ultramicrotomy, depending on their compactness.

Positive or **negative-staining contrast technique** can be sometimes coupled to the fine particle dispersion techniques described above to enhance the contrast of the support of the organic fine particles *via* deposition of heavy-metal salts.

Both casting and ultramicrotomy enable statistical particle analysis, structure determination, the use of contrast enhancement through the negative-staining or decoration-shadowing techniques.

7.6 Inorganic nanocrystal investigation by SEM

Scanning electron microscopy is a very powerful technique to gain morphologic and chemical information on the sample in a fast way. Although its resolution is typically lower than TEM, modern SEM machines are approaching the atomic scale resolution [3]. With respect to TEM, SEM has the advantage to allow working condition at lower acceleration voltage (0.2 keV) thus preventing sample damage. In addition, SEM can be easily exploited to investigate inorganic NCs grown or deposited onto solid support, avoiding time consuming and laborious sample preparation. In this section a brief overview of the impact of the different operative conditions on the image quality of inorganic NC samples will be given.

SEM imaging is strongly affected by several parameters: astigmatism, aperture size, effect of acceleration voltage, conductive properties of the sample. In the following, some examples will be described in order to illustrate how the experimental parameters can be adjusted and optimized as a function of sample properties and requested information.

Acceleration voltage. The acceleration voltage sets the spot size of the beam and electron energy interacting with the specimen, determining the electron penetration volume or depth (interaction volume or signal generation volume), which is defined as the volume in which the electrons spend their energy in signal generating interaction with local atoms. In particular, working at low acceleration voltage results in an increased spot size arising from chromatic aberration in the lenses used to focus the beam. Therefore the higher the acceleration voltage, the smaller the spot size, while at lower energies the electron beam diameter become larger. Conversely, the **penetration depth** of the beam is in direct relation to the acceleration voltage. **Interaction volume simulation** can be carried out to predict the penetration depth depending on the specimen composition and operating acceleration voltage. They rely on Monte Carlo simulations of electron trajectories taking into account the energy of the primary beam electron, the probability of an interaction, the change in direction, energy and mean free path of the electron and a “random” factor for any given interaction.

The dependence of the penetration depth on the acceleration voltage and atomic number is depicted in Figure 7.13. The interaction volume increases with the increase in the acceleration voltage (Fig. 7.13a), while it decreases for high atomic number (Fig. 7.13b).

The higher the acceleration voltage, the higher the energy of electrons and the more deeply the electron beam penetrates inside the specimen, thus making it also less sensitive to the surface.

Moreover, the higher the atomic number of the specimen, the more effectively it can stops or adsorb the electrons, thus resulting in a reduced interaction volume. Penetration depth moves from a pear-like shape for carbon to an apple-like shape for noble metals.

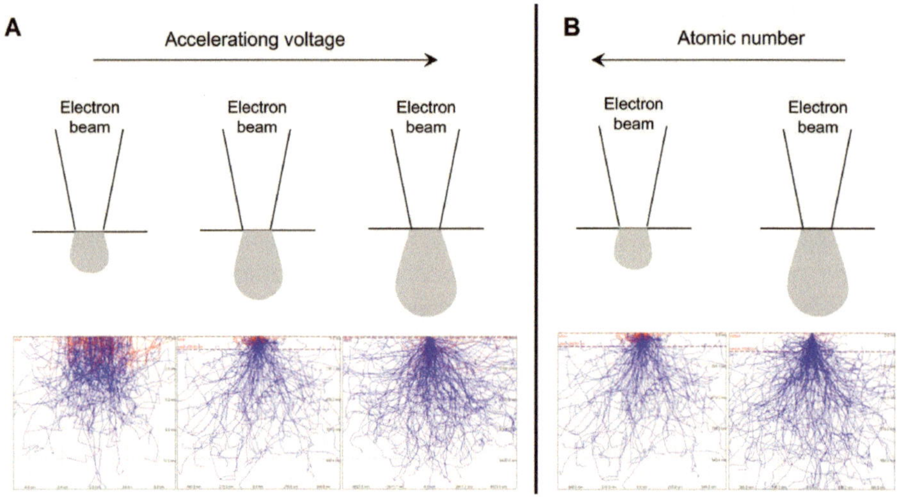

Fig. 7.13: Scheme and Monte Carlo simulation of the dependence of the interaction volume to the acceleration voltage (a) and atomic number of the specimen (b). Simulations were obtained by Casino 2.0 software.

The most suitable value of acceleration voltage depends on specimen composition, morphology and topography, and on the type of information to be achieved. SEM images of colloidal core-shell type NPs, formed of a gold core coated by a silica shell drop cast on a clean silicon substrate, acquired at two different acceleration voltages, are reported in Figure 7.14 as examples [4]. At 1.5 kV the image of the specimen (Fig. 7.14a) reveals the NP surface but not sufficiently to provide the picture of the core-shell structure due to the small penetration volume of the electron. Setting the voltage at 5 kV makes the electron beam penetrate much more deeply in the specimen, thus clearly revealing the presence of the core (Fig. 7.14b). The contrast between shell and core in the NP relies on the different atomic number of the two materials, silica and gold, respectively.

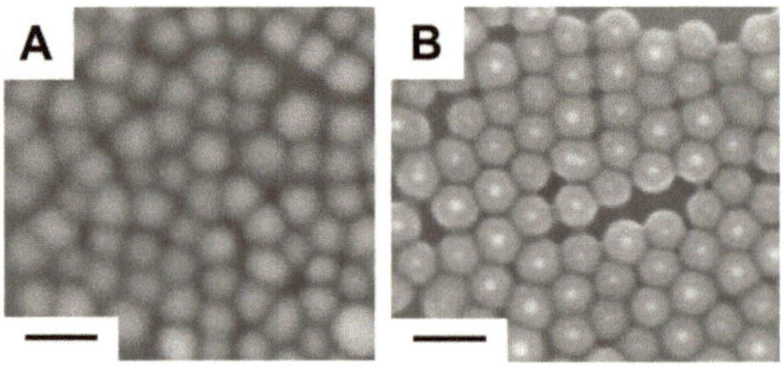

Fig. 7.14: SEM images of Au@SiO2 core-shell type nanoparticles of nearly 40 nm. Aperture size 30 μm; WD 2.3 mm; Acceleration voltage 1.5 kV (a) and 5 kV (b). Scale bar 100 nm [4].

High acceleration voltage could also bring some disadvantages. Scanning a specimen characterized by a fine nanostructured surface at a high acceleration voltage (Fig. 7.15a) [5] makes the contrast enhancement of the fine structures difficult. In fact, as the diffusion volume of the electron beam is large, the BSE produced in such a nanostructured surface generate SE at the specimen surface which increase the background signal. Conversely, a low acceleration voltage (Fig. 7.15b) allows a clearer characterization of the surface fine structure, and higher resolution.

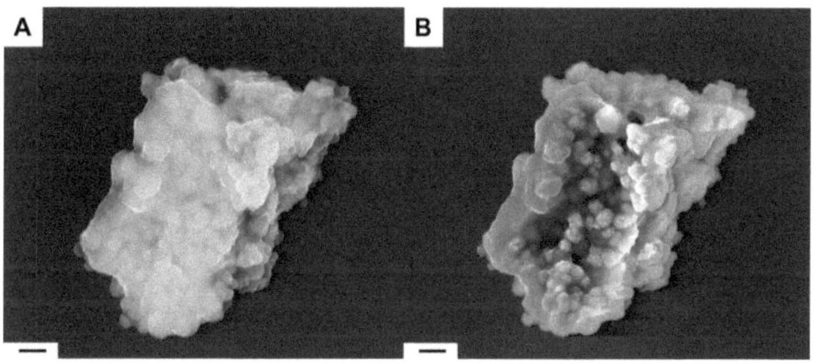

Fig. 7.15: SEM images of TiO_2 NC aggregate at acceleration voltage 15,00kV (a) and 5 kV. Aperture size 20 µm WD 2.3 mm. Scale bar 200 nm [5].

Moreover, at high acceleration voltage, phenomena such as edge-effect and charge-up on the specimen surface are concomitantly increased. The **edge-effect** can be described by considering that that SE emission from the specimen surface strongly depends on the incident angle of the probe at the specimen surface: more SE can leave the sample at its edges leading to increased brightness of these area. The extent of such an edge-effect depends on the acceleration voltage.

In Figure 7.16a the edge-effect on SiO_2 spherical NPs can be clearly detected, as the beads appear much brighter at the edge.

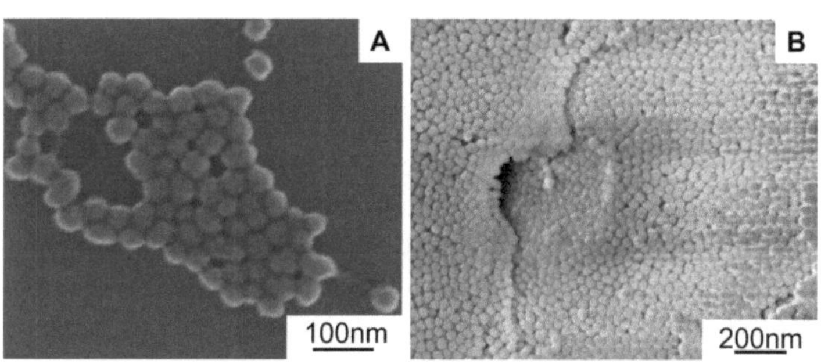

Fig. 7.16: a) Edge effect (acquisition at 10 kV acceleration voltage, 2 mm) and (b) charging effect (acquisition at 2 kV acceleration voltage, 1 mm WD) on SiO_2 beads deposited on silicon slide [4].

The **charging effect** is observed when an excess of electrons is built up on the surface of the sample, and typically occurs when the specimen receives more electrons than those it can emit. Such a "charging" creates an electric field which deflects the electron beam in undesirable ways. In addition, the charge can be randomly released from the sample, exciting the detector in unpredictable ways. Generally, the more insulating the sample is, the higher, and accordingly more critical, the charging effect. Different charging phenomena can be observed: (i) general charging, built up over the active scan area and making the image increasingly bright since contrast and brightness adjustments cannot properly compensate; (ii) edge charging, built up on more relieved or isolated portions of the sample, and causing edges and small features to become unrealistically bright in the image; (iii) area charging, characterized by electron charge and discharge occurring in certain areas of the sample, that become increasingly bright, or alternatively, also darker, during imaging; (iv) line by line charging, generated by electrons randomly released from the sample, line by line, causing very bright streaks across the image; (v) residual charging, generated by electrons left from a previous scan, that add to those emitted by the actual scan. An example of area charging is reported in Figure 7.16b.

Indeed, this charging effect strongly depends on the material composition. If a conducting specimen is covered by a nonconducting coating, the charging effect can be successfully suppressed using an acceleration voltage high enough to let the beam penetrate the coating. Conversely, in the case of a conductive coating sputtered on a nonconductive specimen, low acceleration voltage, which brings small depth of penetration, allows specimen imaging with a reduced charging effect.

Finally, the resolution at low acceleration voltage, although poorer than that obtained using a high acceleration voltage, results in reduced specimen damage.

Aperture size. The role of the objective aperture is to block the straying of electrons, to narrow the beam and thus determine the corresponding convergence angle (α), which is indeed also influenced by the WD. By keeping the WD fixed, it is possible to check the effect of the aperture size on the image quality.

A larger diameter aperture (Fig. 7.17a) results in a larger convergence angle for the beam and thus a reduced depth of focus (which is indicated as the black portion of the cone in Figure 7.17). Similarly, a narrower diameter aperture (Fig. 7.17b) results in an increased depth of focus. On the other hand, the use of a small aperture has two important consequences: an increase in aperture diffraction effects and a decrease in the current reaching the sample (decreases signal to noise ratio). In Figure 7.18, SEM images of a TiO_2 NC aggregate acquired by using three different aperture sizes are reported. The larger aperture (Fig. 7.18a) results in not all the specimen being in focus as the bottom part is not clear. By decreasing the aperture size to 20 nm (Fig. 7.18b) an improvement in specimen imaging is achieved due to a larger depth of focus, all the sample is almost in focus. Finally, a further decrease in aperture size (Fig. 7.18c), which would be expected to bring a larger depth of focus, results in an acquired specimen morphology which is not clear due to the decrease in the signal to noise ratio.

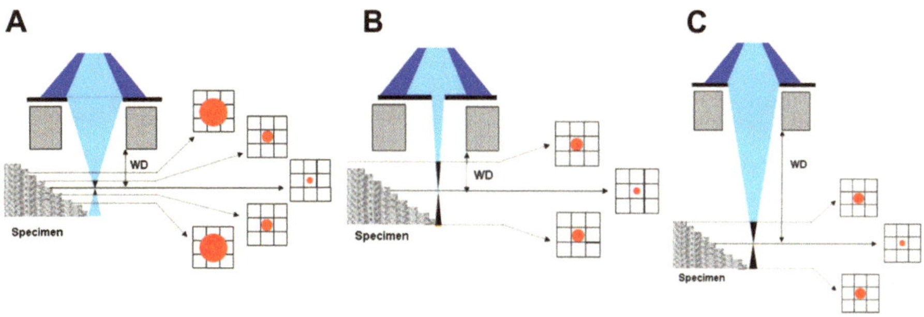

Fig. 7.17: Aperture size (a,b) and WD (c) dependence of the depth of focus.

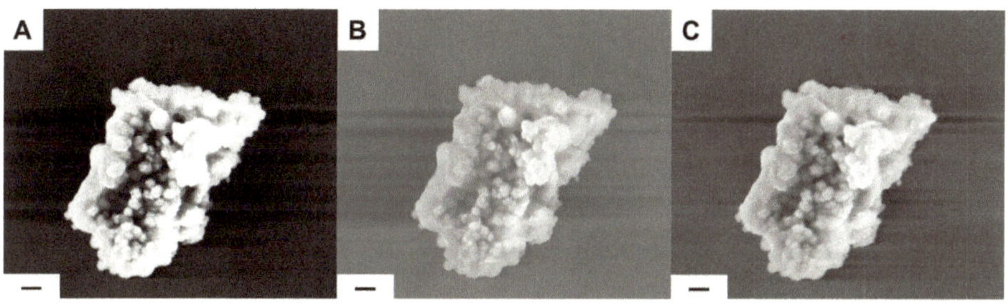

Fig. 7.18: SEM images of TiO_2 NC aggregate with different aperture sizes. Acceleration voltage 5 kV, WD 2.5 mm, magnification 62 Kx, aperture size 30 μm (a) 20 μm (b) 10 μm (c). Scale bar 200 nm [5].

Increasing the WD at a fixed aperture size results in an increase in the depth of focus although it comes to the cost of the attainable resolution.

Focus and astigmatism. One of the most common and relevant issue is represented by **astigmatism and its correction**. Figure 7.19 shows the SEM images of a specimen in focus and out-of-focus, respectively. The images become blurry and unsharp when the objective lens is changed to over-focus at a low magnification. An image is considered astigmatic when it shows unidirectional defocusing moving the objective lens to under or over focus, and a stretching in a particular direction can be observed, as shown in Figure 7.19c.

Scan rate. High quality SEM images require a high signal to noise ratio S/N. This can be obtained by using a high beam current, and a slow scan rate in order to increase the time "spent" by the beam on each pixel, thus allowing to collect a higher amount of electrons along the beam scan. When the signal is weak, an even lower speed is required to improve the signal to noise ratio, but this implies a possible drift in the electron beam, thus affecting the accuracy of the analysis. Thus a balance between electron collection and electron beam drift should be considered in setting the conditions.

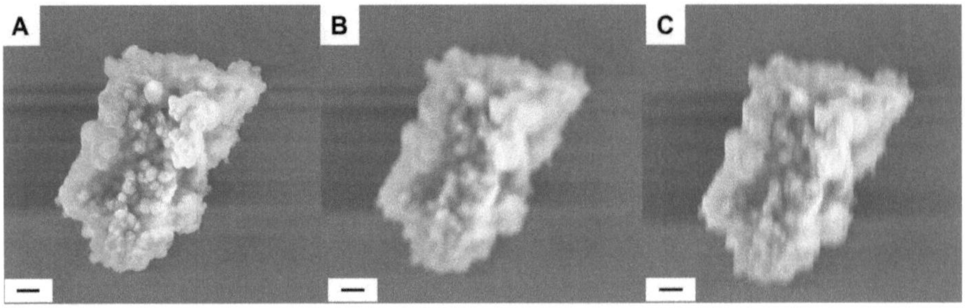

Fig. 7.19: SEM images of TiO_2 NC aggregate in focus (a) out of focus (b) and out of focus with astigmatism (c). Acceleration voltage 5kV, aperture size 20 µm, WD 2.5 mm, mag 62 Kx, in-lens detector. Scale bar 200 µm [5].

Therefore it can be concluded that: resolution can be improved by (i) increasing the intensity of the condenser lens, in order to obtain a sharper electron beam probing the specimen surface, (ii) decreasing the WD and (iii) decreasing the objective aperture. The relationship between resolution and acceleration voltage is complex and depends on the type of information to be achieved. Resolution dependence on the acceleration voltage is related to the fact that voltage affects both the electron beam probe diameter and the interaction volume. An increase in the acceleration voltage certainly results in a smaller probe diameter, which gives a brighter beam and a good contrast that improves resolution. On the other hand, high acceleration voltage increases penetration volume, thus decreasing the overall resolution. If the region of excitation remains small, then the signal will be generated by a small region and no overlap from adjacent regions will take place. Under this condition each individual spot can be resolved from its neighbors. However, even when the beam is scanned in exactly the same positions, or the region of excitation is larger than the regions of signal production, an overlap with the adjacent areas occurs, thus resulting in a poorly resolved image.

Finer surface structure can be generally imaged by using lower acceleration voltages. Since the interaction volume at low voltage is on the same scale as small surface features, these items are much more evident than they would be at higher voltages. At higher acceleration voltages, the deeper beam penetration and increased diffusion area result in signals (*e.g.*, backscattered electrons) generated within the specimen, that are basically not relevant for the surface image, and conversely reduce the contrast in the fine structures.

So, low-voltage SEM can "see" extremely fine (nanometer-scale) features on the surfaces of materials, while high-voltage SEM can see deeper into the material.

Magnification results from the ratio of dimensions of the raster on the specimen and the raster on the cathode ray tube (CRT) of the detector. Assuming that the display size remains fixed, the magnification depends on the sole dimension of the raster on the specimen. This means that SEM magnification can be changed by adjusting the length of the scan on the specimen corresponding to a constant length of scan on the

CRT, which is accomplished by controlling the current supplied to the scan coils of the objective lens.

In Figure 7.20 three images of the same samples at three different magnifications are reported [6]. If the image is set in focus at high magnification, lower magnification images can be easily acquired by simply increasing the scan strength to scan a larger area without losing the focusing condition. This zoom magnification feature is very useful for rapid surveying of the specimen.

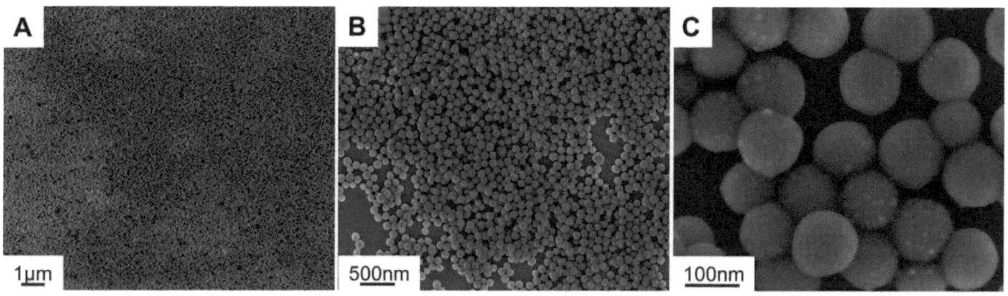

Fig. 7.20: Images of the Au decorated SiO_2 beads at three different magnifications: 11 Kx (a), 40 Kx (b), 320 kx (c). Acceleration voltage 2 kV, WD 2 mm, aperture size 20 μm [6].

Role of the detectors in the imaging. The quality of the images both in terms of resolution and contrast depends on the penetration volume of the electrons generating the signal and on the position of detectors with respect to the specimen.

Noise, blurring, low sharpness, or low contrast in a specimen image can be improved by suitably setting the operating conditions as previously described, but indeed it is surely important to choose the proper detectors according to the specimen surface profile and composition and according to what kind of information we need from the sample.

As reported in Section 7.3, when a primary electron beam hits the sample, secondary (SE) and backscattered electrons (BSE) are re-emitted and can be recorded by means of specific detectors. SEs contain information on sample topography, and can be detected by an **off-lens** detector and by a recently introduced and more efficient **in-lens** detector placed inside the column.

In Figures 7.21 and 7.22 SEM micrographs of nanosized specimen acquired by using an in-lens detector are reported. The high-quality images of the sample, which clearly show the NP shape and size in the range of few nanometers (Fig. 7.21) [7], were recorded using low acceleration voltage (2 kV) and small WD (2.2 mm). These operating condition are surely not suitable for detection by means of an off-lens SE detector.

The improved resolution of the in-lens detector is highlighted in Figure 7.22 which shows composite NPs based on nanosized materials deposited on the surface of another nanoobject, different in composition [8]. The high resolution and good contrast

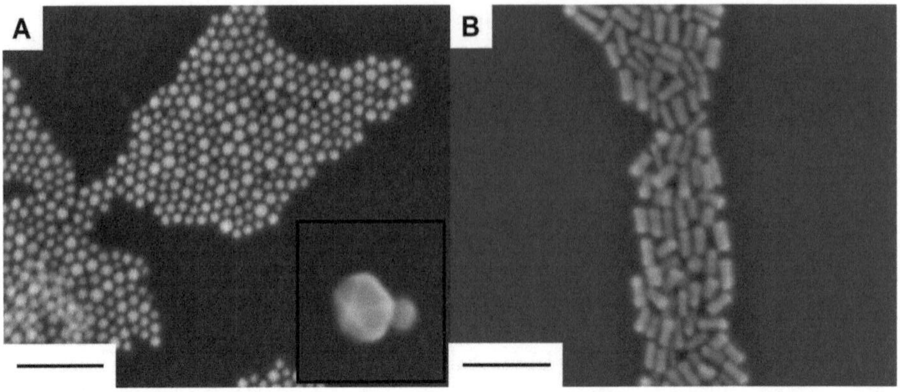

Fig. 7.21: SEM images of Au nanoparticles (a) and Au nanorods (b). Scale bar 100 nm. Acceleration voltage 2 kV, WD 2.2 mm, aperture size 20 μm. In-lens detectors [7].

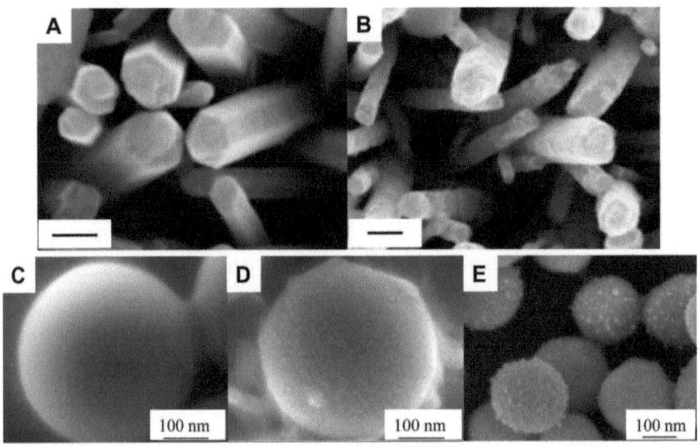

Fig. 7.22: SEM images acquired by in-lens detector of ZnO nanostructures, grown on silicon before (a) and after functionalization with nanocrystals (b). Acceleration voltage 10 kV, WD 2 mm, aperture size 30 μm. Scale bar 100 nm. SEM micrographs with in-lens detector of polystyrene (c) and inorganic nanocrystal decorated polysterene (d) and silica (e) beads. Acceleration voltage 2 kV, WD 4 mm, aperture size 20 μm [8].

provide the characterization of the fine nanostructures and a clear image of the materials.

An increase in the WD, as required for a SE detector, results in a relevant loss of resolution although the acceleration voltage is set to a higher value (Fig. 7.23). The side placement of the SE detector yields images that appear to be naturally lit and three dimensional. Compared to the SE detector, all the other detectors that can be equipped to a SEM microscope, namely the in-lens and BSE detectors, view the sample from above so that the topographic information plays a minor role.

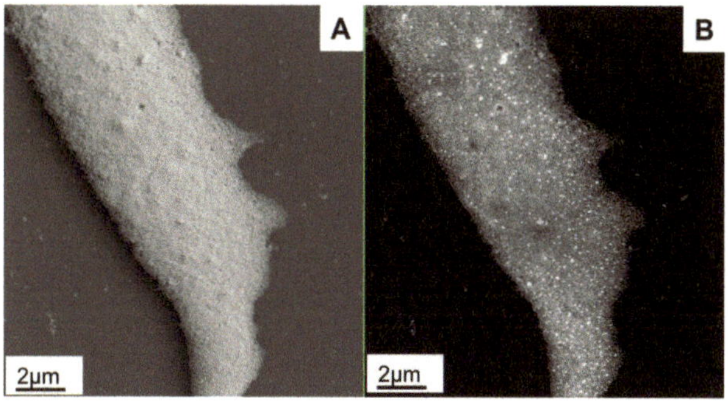

Fig. 7.23: SEM images of Au nanorods on silicon acquired by using off-lens SE (a) and in-lens (b) detector. Acceleration voltage 15 kV, WD 7 mm, aperture size 30 µm [7].

Interestingly, tilting the sample increases the signal in the SE detector and sometimes also the topography, because the tilt axis of the adapted specimen stage is oriented facing the SE detector. Tilting the sample stage (Fig. 7.24) [9] makes the specimen's surface face the detector, which results in a change of the solid angle in which the BSE and SE electrons are emitted from the specimen. As strong tilting of the specimen is disadvantageous for the other detectors, specimens should only be tilted when this is really indispensable.

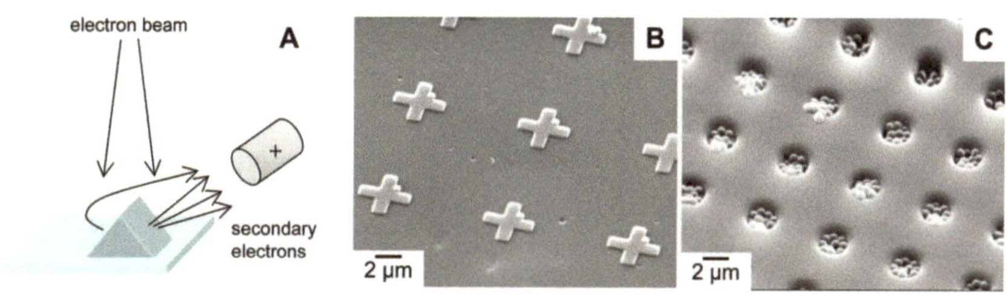

Fig. 7.24: Scheme of SE detection by off-lens detector (a). SEM images of submicrometer-sized nanoparticles assembled on a patterned surface (b,c). Image acquired with the SE detector by tilting the specimen. Magnification 10 Kx, acceleration voltage 15 kV, WD 16 mm. Reprinted with permission from [9]. Copyright 2010 American Chemical Society.

Similarly to the in-lens detector, the BSE detector detects directly inside the beam path and its efficiency decreases considerably when relatively low acceleration (≤ 3 kV) is used. Many applications not only require pure surface information, but also the imaging of material contrasts. In Figure 7.25 images of a TiO_2/Ag NP sample recorded with both the BSE (Fig. 7.25a) and the in-lens SE (Fig. 7.25b) detector are reported [10]. The combination of the two detectors provides topographic and material composition information, which allows a complete and more detailed specimen characterization. In particular, in the BSE image the presence of bright spots also in regions that appear

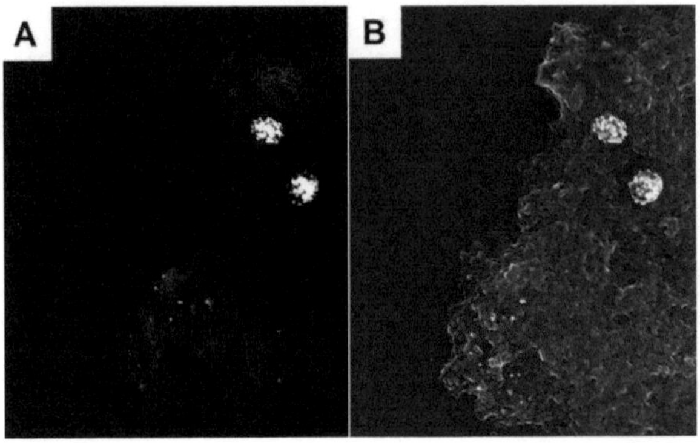

Fig. 7.25: SEM micrographs of TiO_2/Ag nanoparticle specimen acquired by backscattered (a) and in-lens (b) detector. Acceleration voltage 15 kV, ED 2 mm, aperture size 20 µm [10].

dark to in-lens imaging can be noted, suggesting that metallic NP are present inside the TiO_2 aggregate.

7.7 Inorganic nanocrystal investigation by TEM

Nanocrystalline materials have generated a lot of interest in recent years because of their many attractive properties. A modern TEM has the capability to directly image atoms in crystalline specimens at resolutions close to 0.1 nm, smaller than the interatomic distance. An electron beam can also be focused to a diameter smaller than $\sim$ 0.3 nm, allowing quantitative chemical analysis from a single NC. This type of analysis is extremely important for characterizing materials at a length scale from atoms to hundreds of nanometers. Once electrons have interacted with sample they contain information about particle size, shape, crystallinity, and chemical composition. The way we choose which kind of information to get from scattered electrons (*i.e.*, forming an image on the screen or on a CCD camera) is inserting an aperture into the back focal plane of the objective lens, thus blocking out most of the diffraction pattern except that which is visible through the aperture. It is then possible to select the **contrast formation mode.** Contrast modes can be classified in **bright field** (BF), **dark field** (DF) mode and **electron diffraction** mode.

7.7.1 Bright field mode

The most common mode of operation for a TEM is the BF imaging mode. BF TEM image is made up of nonscattered electrons which strike the screen and scattered electrons which thus do not appear as a dark area on the screen. Thicker regions of the sample, or regions with a higher atomic number will appear dark, whilst regions with no sample in the beam path will let the electron striking the screen thus appear bright – hence the term "bright field". The image is in effect assumed to be a simple two-dimensional projection of the sample down the optic axis, and a first approximation may be modeled *via* Beer's law. BF images can give information on size (and size distribution of course), shape, chemical nature of interacting material (only qualitative). In the following, some examples of TEM investigations performed on selected samples of inorganic NCs are reported.

The investigation of the size and geometry of inorganic nanoobjects, along with their size distribution, is a very important issue as it allows to study and, accordingly, to exploit the relevant size- and shape-dependent properties of NCs.

A suitable correlation of NCs' structure and morphology with their preparative conditions allow to elucidate mechanisms behind the growth of inorganic NCs and nanoobjects and, finally, enable an effective *a priori* design of the synthesis of new materials with unprecedented properties. As an example, the investigation of the mechanism underlying the growth of anisotropic Au NPs, namely nanorods (NRs), has been a challenging issue in materials science for two decades. Au NRs are very interesting candidates for several applications, in biomedical, optoelectronic and sensing fields, thanks to their extremely sensitive longitudinal plasmon band which is tunable from the visible to the near infrared range according to NR aspect ratio (*i.e.*, the ratio between length and diameter).

Thanks to a methodical investigation of the size/shape evolution as a function of reaction conditions in combination with convergent beam electron diffraction, a reasonable mechanism accounting for the growth of Au NRs has been proposed [11]. Figure 7.26 reports an example of the study of the evolution of NP size/shape as function of reaction conditions carried out by TEM imaging recorded in BF mode.

In BF imaging mode the image brightness is a function of the atomic number of the interacting element; the image being darker in the presence of heavy atoms. In this perspective this mode can be very useful to image nanocrystalline heterostructured nanocomposite with distinct chemical domains. By using this mode, NPs composed of different chemical domains will be imaged in different shades of grey from lighter to darker according to the interacting element's atomic number. Such a feature can be exploited to distinguish the chemical domain composing hybrid NCs. This is an important class of inorganic nanostructures characterized by a topologically controlled chemical composition. Such hybrid NCs consist of two or more chemically different material sections, that are interconnected through permanent inorganic interfaces and allow the properties of the different materials forming the distinct chemical domains to merge in one structure, thus resulting in a new material with original chemical and physical properties deriving from the fusion of luminescent, magnetic and catalytic material domains in one NP, have been reported (Fig. 7.27) [12–15].

BF TEM imaging can be also very powerful in investigating NC superlattices.

The fabrication of highly-ordered and large-area architectures based on inorganic NCs is a current challenge due to the huge potential in terms of collective properties

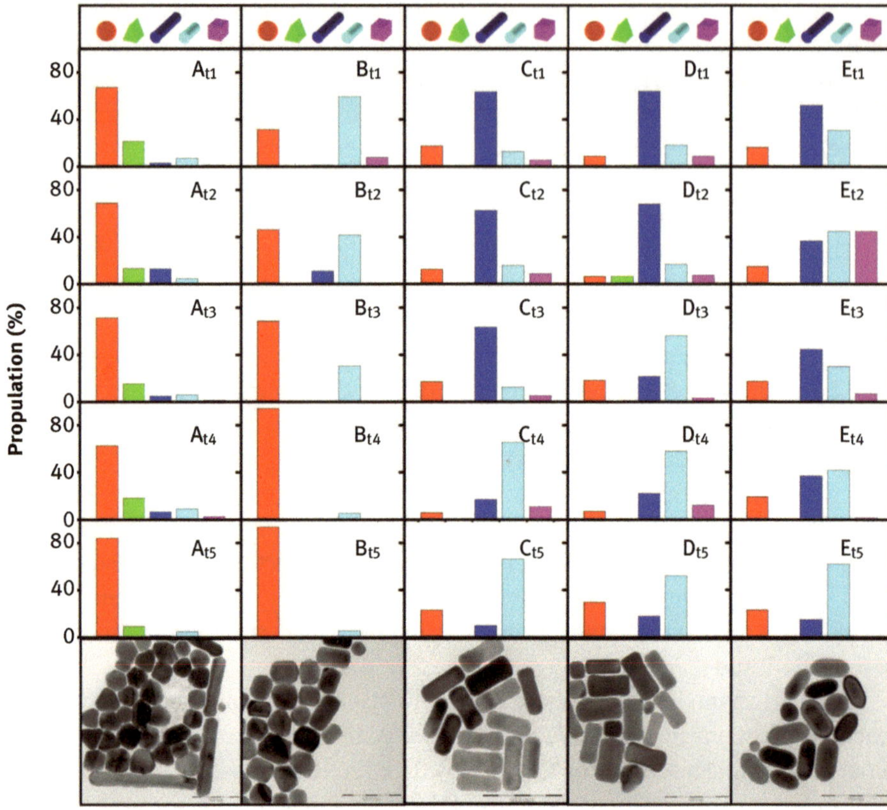

Fig. 7.26: Statistical analysis of Au NP shape evolution as a function of the irradiation time and amount of silver ion used to catalyze the growth of nanorods. The sample sets correspond to: (a) Au NP growth in absence of Ag ions, (b–e) Au NPs grown in the presence of increasing amounts of Ag ions (Au/Ag ratio: 60–15). The irradiation times at which the aliquots have been extracted are denoted in the relevant panels as follows: (t1) 17, (t2) 21, (t3) 26, (t4) 63, (t5) 88 h. The colors in the histograms are related to the main types of shapes observed, as indicated by the legend in the top part of the figure. In the bottom part of the figure, representative TEM images of NPs obtained after irradiating for 17 h in the respective cases are shown (scale bar 100 nm). Reprinted with permission from [11]. Copyright 2009 American Chemical Society.

which could arise from such assemblies, essential in fundamental studies, as well as for applications in advanced technological fields. Interestingly, the collective properties that arise are not just a function of NC chemical composition but also of their spatial arrangement. Thus the investigation of the mechanisms beyond the formation of such NC-based superlattices is of paramount importance [16, 17]. BF TEM imaging is a very powerful tool to investigate the way NCs organize themselves and the parameters driving the final arrangement of the superlattice (NC size, chemical composition, capping agent and so on, Fig. 7.28).

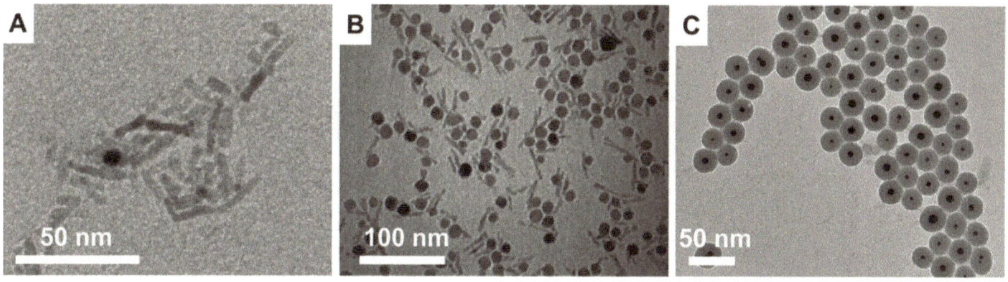

Fig. 7.27: (a) BF image of TiO_2 NR – Ag NP. Reprinted with permission from [14]. Copyright 2011 American Chemical Society. (b) BF image of TiO_2 NR/Fe3O4 NC. Reprinted with permission from [13]. Copyright 2011 American Chemical Society. (c) BF image of Au@SiO2 NPs [15].

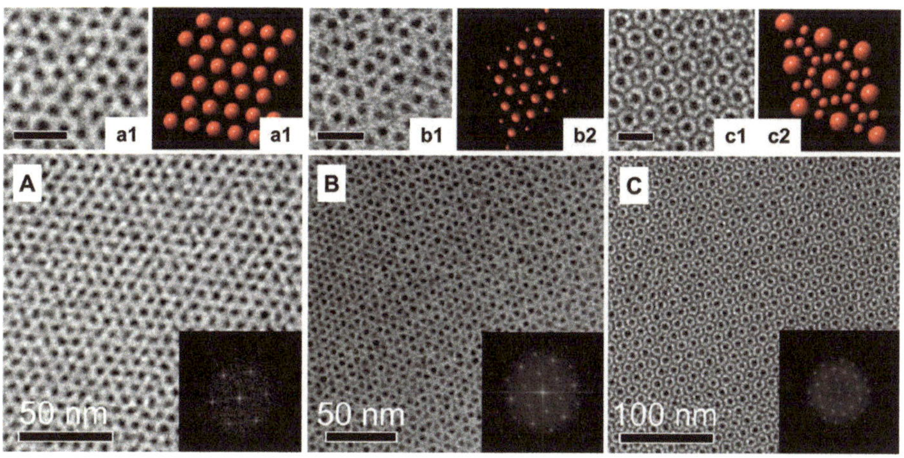

Fig. 7.28: (a) TEM images of 3.9 ± 0.5 nm MSD PbS NCs, (b) of 1.9 ± 0.5 nm and 4.1 ± 0.5 nm BSD PbS NCs, and (c) of 1.9 ± 0.5 nm and 5.4 ± 0.5 nm BSD PbS NCs in toluene solutions, with the corresponding close up on the geometry (scale bar 20 nm), the FFT and the sketch of the assembled geometry. Reprinted with permission from [16]. Copyright 2010 Royal Society of Chemistry.

7.7.2 Dark field contrast mode

With a suitable positioning of apertures in the back focal plane, *i.e.*, the objective aperture, the desired Bragg reflections can be selected (or excluded), thus only the portion of the sample that is causing the electrons to scatter to the selected reflections will be projected onto the imaging apparatus (a fluorescent screen or a CCD camera).

When no scattering occurs from the selected peak of the sample, dark regions form in the image. Therefore, regions without scattering (no material) will appear dark while selected scattered electrons, coming from the sample, will strike the screen producing bright spots. This is known as a **DF image**. A particular kind of DF imaging is annular dark field (ADF, see Section 7.4) imaging which is a method of mapping samples in a STEM. These images are formed by collecting scattered electrons with an annular dark field detector. In traditional DF imaging, only electrons scattering through the objective aperture are collected. Conversely, an ADF detector collects electrons from an annulus around the beam, sampling far more scattered electrons than can pass through the objective aperture. This gives an advantage in terms of signal collection. An ADF image formed only by electrons incoherently scattered at a very high angle – as opposed to Bragg scattered electrons – is highly sensitive to variations in the atomic number of the atoms in the sample (Z-contrast images). This technique is also known as **high-angle annular dark field imaging** (**HAADF**). Typical applications of DF imaging mode are grain size determination or detection of second-phase particles (Fig. 7.29) [18, 19].

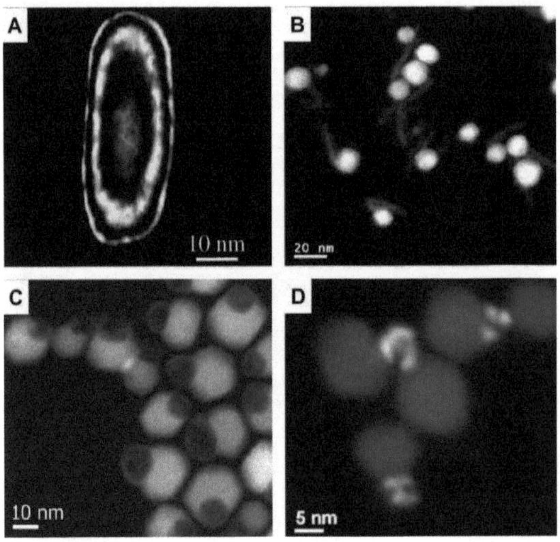

Fig. 7.29: (a) Dark-field TEM image of an Au rod oriented nearly parallel to [110]. Reprinted with permission from [18]. Copyright 2000 American Chemical Society. (b) Low-magnification HAADF image of a group of Fe_3O_4/TiO_2 NR hybrid nanocrystals. Reprinted with permission from [13]. Copyright 2006 American Chemical Society. (c) High-angle annular dark field (HAADF) STEM images of iron oxide nanocontainers and (d) HAADF-STEM image showing the encasement of Pt domains in Fe_3O_4 nanocontainers. Reprinted with permission from [19]. Copyright 2011 American Chemical Society.

7.7.3 Diffraction mode – electron diffraction

In diffraction mode, electrons diffracted by the specimen that are parallel to one another come into focus at the back focal plane of the objective lens.

To operate the TEM in diffraction mode, the objective aperture is removed from the beam path and the microscope is adjusted to focus an image of the back focal plane of the objective lens, not the image plane. The result is an electron diffraction (ED) pattern that is completely dependent on the d-spacing and structure of the crystal under investigation.

7.7.3.1 Selected area diffraction

When electron diffraction is obtained from many different crystals with different orientation, either at the Bragg angle or not, a pattern with well-defined concentric rings, rather than distinct spots, will result.

When electron diffraction is from an amorphous structure (*i.e.*, no crystalline structure) then only a central bright spot is obtained, formed of transmitted electrons and a single ring of randomly forward scattered electrons.

The diffraction pattern is formed of electrons coming from the whole area of the specimen that has been illuminated by the beam. Such a pattern is not very useful because the sample and the viewing screen can be damaged due to the strong intensity of the beam. Therefore, a defined area of the specimen characterized by a high density of NCs needs to be selected in order to form the diffraction pattern. Typically, an aperture can be selected and placed in the image plane of the objective lens, in order to create a virtual aperture at the plane of the specimen. In this way the so-called selected area diffraction (SAD) can be obtained.

For thin crystalline samples, an image consisting of a pattern of dots, in the case of a single crystal, or a series of rings, in the case of a polycrystalline sample or randomly oriented NCs, can be obtained (Fig. 7.30) [20].

Fig. 7.30: Selected area diffraction pattern of Au NPs [20].

7.7.3.2 Convergent beam electron diffraction

Another possibility to select a portion of the sample to record electron diffraction consists in reducing the size of the beam using the condenser lenses in order to focus the beam at the specimen and forming **convergent beam electron diffraction (CBED)** patterns. Converging the beam destroys any coherence and the spots in the pattern are not sharply defined but spread into disks.

For instance, by coupling information from CBED patterns and BF images of the same portion of the sample it is possible to gain information on the orientation of a crystalline plane or on the preferred direction of elongation of the selected NCs (Fig. 7.31) [11].

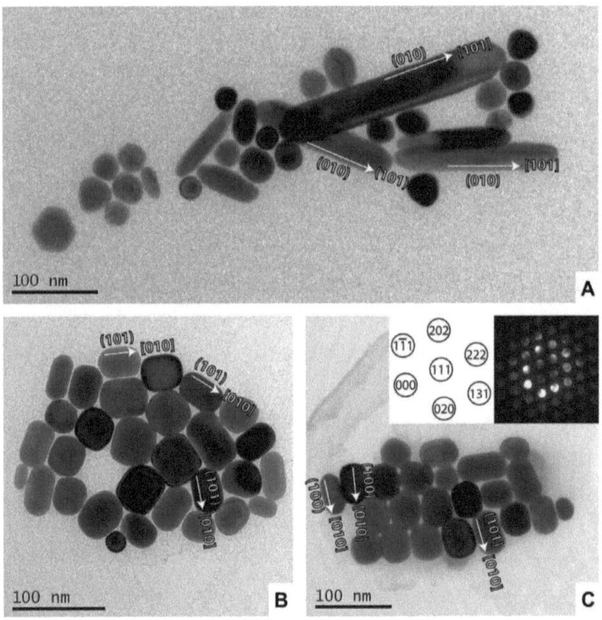

Fig. 7.31: Bright-field TEM images of small groups of AuNRs: (a) NRs prepared in the absence and (b and c) in the presence of Ag ions. Preferred elongation directions and longitudinal side facets are indicated. A typical CBED pattern (along with the indexation scheme for fcc gold) relative to a NR down its {110} zone axis, which has been used to determine the structural features of the NPs, is also shown in the inset of panel (c). Reprinted with permission from [11]. Copyright 2009 American Chemical Society.

7.7.3.3 Investigating crystalline structure: High-resolution TEM

High-resolution transmission electron microscopy (HRTEM) is an imaging mode of TEM that allows the imaging of the crystallographic structure of a sample at an atomic scale. Because of its high resolution, it is a valuable tool to study nanoscale properties of crystalline materials such as semiconductors and metals. A high-resolution image is an interference image of the transmitted and the diffracted beams. By operating the microscope in high resolution mode individual atoms, crystalline defects and different crystalline domains can be imaged (Fig. 7.32) [14, 21].

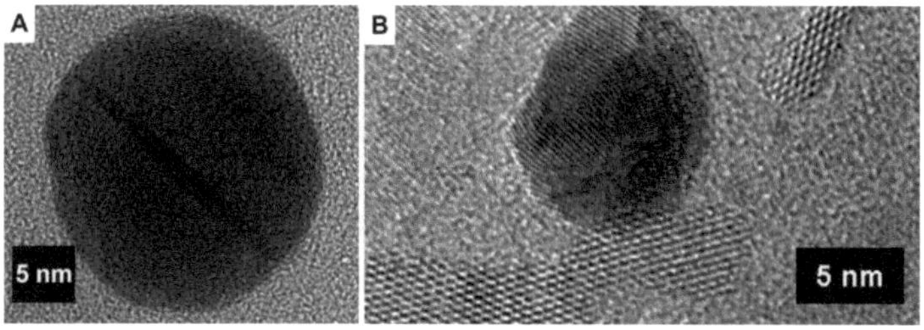

Fig. 7.32: (a) High-resolution TEM image of a single Au NP (a twin plane is visible). Reprinted with permission from [21]. Copyright 2006 WILEY-VCH Verlag GmbH & Co. (b) High-resolution detail of an Ag particle directly in contact with a TiO_2 NRs. Reprinted with permission from [14]. Copyright 2009 American Chemical Society.

7.8 Chemical analysis by electron microscopy

Electrons scattered by the sample contain information about its topography, its structure and its **chemical composition**. Indeed, according to differing scattering events (elastic or inelastic) different information can be obtained about the sample. Several different techniques have been developed in the last decades in order to exploit scattered electrons to gain qualitative and quantitative information on the chemical composition of the sample.

Under the impact of an incident electron, the electrons bounded to the atoms may be excited either to a free electron state or to an unoccupied energy level with a higher energy. The quantum transitions associated with these excitations will emit photons (*e.g.*, X-rays) and Auger electrons. These inelastic scattering signals are the fingerprints of the elements that can provide quantitative chemical and electronic structural information.

7.8.1 Energy dispersion spectroscopy (EDS)

EDS (also known as energy dispersive X-ray spectroscopy EDX) is mainly sensitive to heavy elements due to the fluorescence effect arising from Auger emission for light elements. X-rays are indirectly produced when an electron is displaced through a collision with a primary beam electron and is replaced by another electron. The resultant loss of energy is given off in the form of an X-ray. The energy will always be less than the energy of the primary beam electron.

Each element has characteristic peak positions corresponding to the possible transitions in its electron shell. For instance, copper is characterized by two K-peaks at about 8.0 and 8.9 keV and an L-peak at 0.85 eV.

In heavy elements like tungsten, a lot of different transitions are possible and many peaks are then present.

EDS can be implemented on both SEM and TEM. In particular, when implemented on machines working in scanning mode EDS can provide useful chemical mappings allowing the construction of a false color image in which each color is associated with a specific chemical element.

7.8.2 Electron energy loss spectroscopy (EELS)

Electrons undergoing inelastic scattering with the sample will lose energy and have their paths slightly and randomly deflected. The amount of energy loss can be measured *via* an electron spectrometer and interpreted in terms of what caused the energy loss. Since electron energy loss spectroscopy (EELS) requires thin samples for detection, it can only be implemented on TEM machines.

A typical EELS spectrum consists of three parts: (1) **Zero-loss peak** at 0 eV: This mainly contains electrons that still have the original beam energy E_0, *i.e.*,, they have only interacted elastically or not at all with the specimen. In thin specimens, the intensity of the zero-loss beam is high, so that damage of the CCD chip can occur. Since there is no useful information in it, the zero-loss beam is often omitted during spectrum collection. (2) **Low-loss region** (< 100 eV): Here, the electrons that have induced plasmon oscillations occur. Since the plasmon generation is the most frequent inelastic interaction of electrons with the sample, the intensity in this region is relatively high. Intensity and number of plasmon peaks increases with specimen thickness. (3) **High-loss region** (> 100 eV): For the ionization of atoms, a specific minimum energy, the critical ionization energy E_C or ionization threshold, must be transferred from the incident electron to the expelled inner-shell electron, which leads to ionization edges in the spectrum at energy losses that are characteristic for an element.

EELS is complementary to X-ray spectroscopy, and it can be utilized for qualitative and quantitative element analysis as well. In particular, the detection of light elements is a main task of EELS. Compared to plasmon generation, the inner-shell ionization is a much less probable process, leading to a low intensity of the peaks. In the high-loss region, the amount of inelastically scattered electrons drastically decreases with increasing energy loss, thus small peaks are superimposed on a strongly decreasing background. Because of the low intensity, the representation of the high-loss region is often strongly enhanced.

The critical ionization energy E_C is sensitive to the chemical status of the element: *e.g.*, the L-edge of Cu metal and CuO are shifted with respect to each other (chemical shift). Moreover, the ionization process may take more energy than E_C, and therefore signal intensity is still present even after the corresponding edge. Actually, this region, designated as ELNES (energy-loss near-edge structure), mirrors the density of states and provides information about the chemical binding. Modulations further away from the ionization edge contain information about inter-atomic distances and coordination (EXELFS, extended energy-loss fine structure).

A lot of information is present in an EEL spectrum. The energy resolution (below 1 eV) is much higher than in X-ray spectroscopy, and as a result much more structural detail can be obtained from the fine structure in EELS. For many questions, it is important to get this information with a high local resolution. Such a mapping can be done by two different methods:

- **STEM with EELS.** In STEM mode, a region of the sample is selected, and an EELS is measured on each spot of the defined grid in order to obtain a chemical mapping of the sample (serial detection).
- **EFTEM.** Using an energy filter an energy range is selected, and an energy-filtered image is recorded with electrons of this energy (parallel detection). Such methods will be discussed in detail in the next section.

7.8.3 Energy-filtered transmission electron microscopy (EFTEM)

Energy filtering is a valuable tool for materials characterization. It can be applied for solving several questions and elucidate relevant issues. In a zero-loss filtering approach, that is to say when only the zero-loss beam is considered, all inelastically scattered electron are omitted (reduction of noise). This leads to increased contrast in TEM images (important for specimens with low contrast, *e.g.*, biological specimens) or in diffraction patterns.

Energy filtering can also be exploited for electron spectroscopic imaging. By using electrons with a well-defined energy loss (ionization edge), elemental distribution maps can be generated.

Typically an energy filter is mounted below the column of a TEM. In the curved magnetic field, electrons with different energies are dispersed, and consequently an EEL spectrum is obtained at the end of this part. In fact, the magnetic prism is the same as in a parallel electron energy loss spectrometer, and, of course, an EEL spectrum can be measured with a suitable filter as well as with an optical column, in which an energy range is selected by a slit located at the dispersion plane. In imaging mode, electrons of the selected energy pass through the lenses of the optical system, and an energy-filtered image occurs at the end of the filter, where a slow-scan CCD camera is installed for image recording.

7.8.4 Some examples of chemical analysis in electron microscopy

Three analytical techniques are most commonly used in chemical microanalysis in EM: EDS, EELS and EFTEM. In particular, EDS can be implemented on both SEM and TEM whereas EELS and EFTEM work only on very thin samples and thus only on TEM samples.

The size of the incident electron probe can be as small as 2–3 nm, allowing to acquire EDS and EELS spectra from individual NCs. The data can be used quantitatively for determining the chemical compositions of NPs. EDS is mainly sensitive to heavy elements due to the fluorescence effect arising from Auger emission for light elements, while EELS is most adequate for light elements. Both EDS and EELS can be applied complementarily to determine the chemical composition. The detection limit for EDS can be as high as 1 %, and that for EELS can be as high as 5 %, depending on elements [18].

Characteristic X-rays are useful for microchemical analysis by dispersive X-ray spectrometry (EDS). An EDS spectrum contains peaks at the energies of the characteristic X-rays from the elements in the material.

Chemical mapping of element distributions in samples is also possible when EELS is performed in STEM mode. The electron beam is focused into a small probe and EELS spectra are acquired from a two-dimensional grid of points across the sample. Each "pixel" in the image can contain an entire EELS spectrum. The data set contains information on chemical variations across the specimen (Fig. 7.33) [22, 23].

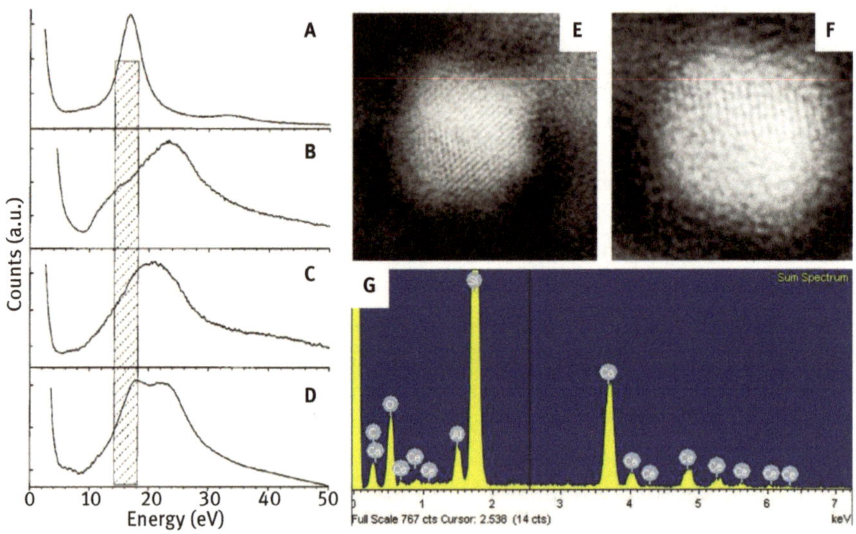

Fig. 7.33: Typical EELS spectra relative to crystalline silicon (a), amorphous SiO_2 (b), SiO_2 thin film obtained by plasma enhanced chemical vapor deposition (c), the same SiO_2 film after a thermal annealing at 1250 °C (d). For all spectra, the zero-loss peak is not fully plotted. The energy window used to generate the EFTEM images in (e) and (f) is reported. Reprinted with permission from [22]. Copyright 2004 American Institute of Physics. (e,f) High-resolution EFTEM images of Si NC. Note, in (e) the presence of an amorphous shell, about 1 nm thick, covering the crystalline core of the cluster. In (f) note the complete absence of reticular planes in half of the Si nanocluster. Reprinted with permission from [22]. Copyright 2004 American Institute of Physics. (g) EDS spectrum recorded by SEM of a mixed oxide nanocrystalline sample composed of CeO, CaO, and Al_2O_3 deposited onto silicon [22].

7.9 Conclusions

The objective of this chapter was to introduce the reader to inorganic nanomaterial characterization by electron microscopy which is a very powerful tool to characterize inorganic nanocrystals. Controlling the shape of nanocrystals is the state of the art in nanophase materials research. The shape-controlled nanocrystals have many prospects that are likely to impact the fields of catalysis, self-assembly, and nanodevices. Modern electron microscopes allow atomic scale resolution allowing to understand the behavior of matter at the nanoscale. The working principle of SEM and TEM have discussed as well as the operative conditions to optimize the quality of images. Sample preparation plays an important role in order to obtain a thin slide of the samples avoiding artifacts thus allowing to obtain the desired information from the sample.

The advantage of electron microscopy is that once electrons have interacted with the sample they contain a range of information on topography (*i.e.*, surface features of an object), morphology (*i.e.*, size and shape of nanoparticles), structure (*i.e.*, how atoms are arranged in the object), composition (relative amount of elements and compounds making up the object). By suitably choosing the operative conditions it is possible to select the desired information to be analyzed.

Bibliography

[1] Echlin P. Handbook of Sample Preparation for Scanning Electron Microscopy and X-Ray Microanalysis: Springer US; 2009.
[2] Ayache J, Beaunier L, Boumendil J, Ehret G, Laub D. Sample Preparation Handbook for Transmission Electron Microscopy – Methodology: Springer New York; 2010.
[3] Obayashi H. In celebration of the 60th anniversary of Journal of Electron Microscopy. Journal of Electron Microscopy, 2011; 60: S283–S6.
[4] Fanizza E, Depalo N, Comparelli R, Striccoli M, Curri ML. Unpublished data. In; 2012.
[5] Comparelli R, Fanizza E, Striccoli M, Curri ML. Unpublished data. In; 2012.
[6] Corricelli M, Fanizza E, Comparelli R, Striccoli M, Curri ML. Unpublished data. 2012.
[7] Placido T, Fanizza E, Comparelli R, Striccoli M, Curri ML. Unpublished data. 2012.
[8] Binetti E, Fanizza E, Comparelli R, Striccoli M, Curri ML. Unpublished data. 2012.
[9] Fanizza E, Malaquin L, Kraus T, et al. Precision patterning with luminescent nanocrystal-functionalized beads. Langmuir, 2010; 26: 14294–300.
[10] Petronella F, Comparelli R, Fanizza E, Striccoli M, Curri ML. Unpublished data. 2012.
[11] Placido T, Comparelli R, Giannici F, et al. Photochemical synthesis of water-soluble gold nanorods: The role of silver in assisting anisotropic growth. Chem Mater, 2009; 21: 4192–202.
[12] Casavola M, Buonsanti R, Caputo G, Cozzoli PD. Colloidal strategies for preparing oxide-based hybrid nanocrystals. Eur J Inorg Chem 2008; 2008: 837–54.
[13] Buonsanti R, Grillo V, Carlino E, et al. Seeded growth of asymmetric binary nanocrystals made of a semiconductor TiO_2 rodlike section and a magnetic g-Fe_2O_3 spherical domain. J Am Chem Soc 2006; 128: 16953–70.

[14] Petronella F, Fanizza E, Mascolo G, et al. Photocatalytic activity of nanocomposite catalyst films based on nanocrystalline metal/semiconductors. J Phys Chem C 2011; 115: 12033–40.
[15] Depalo N, Fanizza E, Comparelli R, Striccoli M, Curri ML. Unpublished data. 2012.
[16] Corricelli M, Striccoli M, Comparelli R, Curri ML. Colloidal chemistry routes for fabrication of nanoparticle-based metamaterials, in: Johnson NP, Ozbay E, Ziolkowski RW, Zheludev NI (eds.), Proc. SPIE 7711, Metamaterials V, (29 April 2010), p. 77111A SPIE, Brussels, 2010.
[17] Altamura D, De Caro L, Corricelli M, et al. Meso-crystallographic study of a three-dimensional self-assembled bimodal nanocrystal superlattice. Cryst Growth Des 2012; 12: 1970–6.
[18] Wang ZL. Transmission electron microscopy of shape-controlled nanocrystals and their assemblies. J Phys Chem B 2000; 104: 1153–75.
[19] George C, Dorfs D, Bertoni G, et al. A cast-mold approach to iron oxide and Pt/iron oxide nanocontainers and nanoparticles with a reactive concave surface. J Am Chem Soc 2011; 133: 2205–17.
[20] Capitani G, Placido T, Comparelli R, Striccoli M, Curri ML. Unpublished data. 2012.
[21] Cozzoli PD, Curri ML, Giannini C, Agostiano A. Synthesis of TiO2-Au composites by titania-N anorod-assisted generation of gold nanoparticles at aqueous/nonpolar interfaces. Small 2006; 2: 413–21.
[22] Iacona F, Bongiorno C, Spinella C, Boninelli S, Priolo F. Formation and evolution of luminescent Si nanoclusters produced by thermal annealing of SiO[sub x] films. J Appl Phys 2004; 95: 3723–32.
[23] Comparelli R, Pastore C, Dibenedetto A, Aresta M. Unpublished data. 2012.

Angela Dibenedetto, Stefania Fasciano, Antonella Colucci, and Michele Aresta

8 Nanosized particles: questioned for their potential toxicity, but some are applied in biomedicine

8.1 Introduction

Particles having a dimension smaller than 100 nm are labeled “nanoparticles” and have properties that are not shared by nonnanoscale materials [1]. The exposure of humans to airborne nanoparticles has exponentially grown in the last century. The recent development of “nanoscience and nanotechnology” has largely increased the number of “engineered nanoparticles” that find application in an ever growing number of cases. The peculiar “surface structure” influences the “reactivity” of nanoparticles and makes them potentially quite “aggressive” towards biosystems. This is raising a lot of concern about the potential toxicity of nanoparticles, considering that they can be taken up more easily than nonnanoscale materials and reach the spleen, bone marrow and the nervous system, organs, tissue and cells, causing stress, DNA mutation and may induce structural damage in mitochondria, with cell death. Such fears are pushing towards the definition of potential dangers that nanoparticles may cause to the human organism, increasing precautions in their manufacture and management, accuracy in their emission measurement, increasing the knowledge of the routes of exposure and up-take and of the mechanism of action on organs, tissues and cells, and controlling the end of life disposal, in order to make sure and safe the production and use of such materials that also find application in biomedicine for a variety of uses.

8.2 Nanoparticles classification

The dynamic development of nanoscience and nanotechnology during the last 20 years has brought the production of a large number of new nanomaterials [2, 3] that now count more than 1000 new products in use with a total mass of some 1200 tons. The forecast is that by 2014 $> 15\,\%$ of the products on the global market may carry a nanocomponent or be based on nanotechnology for their production [4] with great uncertainty about the potential impact on human health and the environment.

Nanoparticles have dimensions in the range 1–100 nm [2] and may exist as 1D (deposited layer on a solid surface), 2D (nanowires or nanotubes attached to a large surface, or nanopores in a solid surface or membrane) and 3D objects (“single” particles, such as wires, nanotubes, spheres, etc.). When the nanoparticles are bound to a macrostructure or incorporated in it, the risk of use is lower than when they are used as a single particle, but still the risk associated to their production remains quite non-

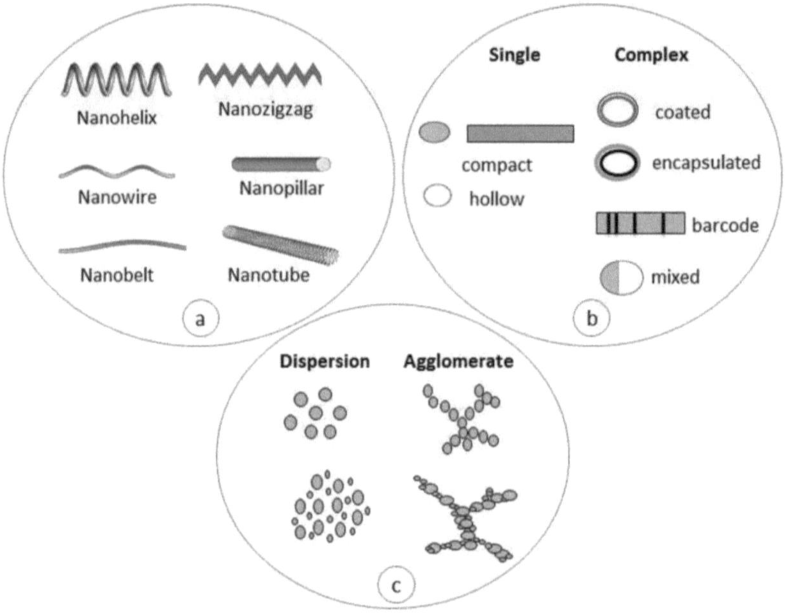

Fig. 8.1: Classification of nanoparticles based on: (a) the morphological aspect, (b) the composition, and (c) the homogeneous or non-homogeneous state.

defined. Figure 8.1 shows some basic information about nanoparticles, emphasizing factors that may influence their bio-activity.

The structural characterization of nanoparticles thus plays a key role in understanding their properties and finding a structure-activity correlation. Nanoparticles may have various compositions, encompassing purely inorganic, hybrid inorganic-organic, and organic materials. Inorganic nanoparticles owe their activity to the large number of atoms per unit surface. It is noteworthy that, although the definition of nanoparticle includes sizes from 1 to 100 nm, nevertheless the properties may considerably change with size as shown in Figure 8.2.

Figure 8.2 shows that the number of atoms on the surface very rapidly increases in the range from 1–5 nm. The total interval can be quite correctly divided into 5 parts, identifying the ranges 1–5 nm, 6–15 nm, 16–30 nm, 31–50 nm, 51–100 nm as the regions with different properties and activity. Nanoparticles with a dimension smaller than 15 nm may have quite unique properties due to their particular thermodynamics [5]. Such nanoparticles may be stabilized by changes in their crystallographic structure [6]. Physical properties that may be influenced by the nanodimensions are: transition temperatures [7], magnetic and electric properties [8], phase transition (the melting temperature can be decreased also by more than 100 °C passing from bulk- to nanomaterials) [9] and dissolution. Such effects can be predicted from thermodynamics (Laplace equation) and particles with dimensions below 15 nm behave quite differ-

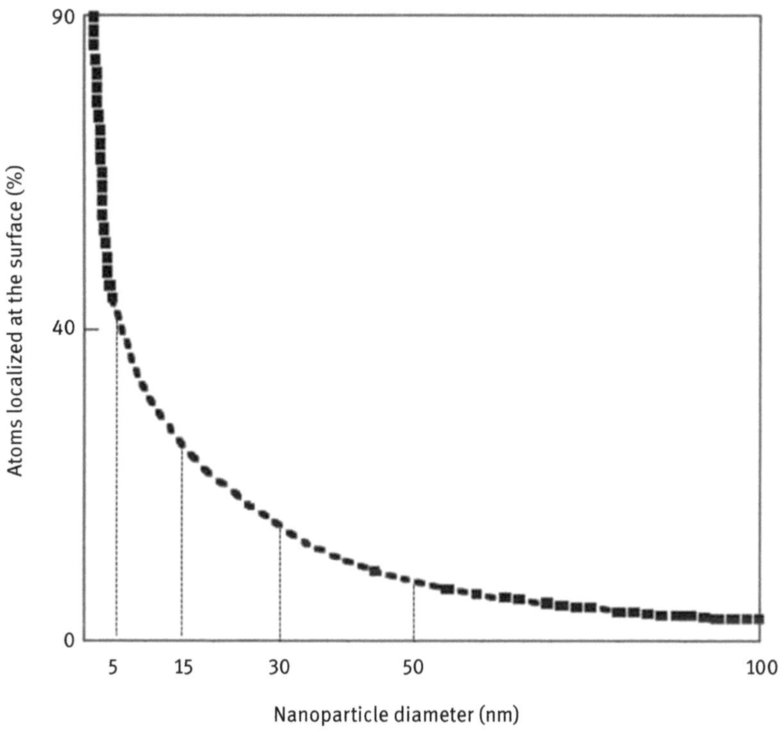

Fig. 8.2: The number of atoms on the surface (%) changes with the nanoparticle diameter (nm).

ently from those that have a dimension over 30 nm. Several particular properties depend on the change of surface tension, γ, and on the derivative $d\gamma/dr$ that is used in thermodynamic models [10].

8.3 Nanoparticles and biosystems

Several well-documented important changes occur when downscaling the dimension of particles. A quite well-known case is the change in the catalytic properties of gold. It is well known that bulk gold is inactive (noble metal) but nanoparticles of < 2 nm are very active in air-oxidation catalysis [11]. The implication of gold nanoparticles has been quite clearly demonstrated. This achievement is of great importance as it opens up a new technology, for example in the oxidation of olefins and, consequently, to polycarbonates obtained by direct carboxylation of epoxides. The use of air instead of hydrogen peroxide would foster the production of large volume applications now limited by the availability of H_2O_2 [12].

Also the TiO_2 photocatalytic activity is largely influenced by the dimension of the particles. In fact, nanosized particles in the range 7–25 nm find better application in

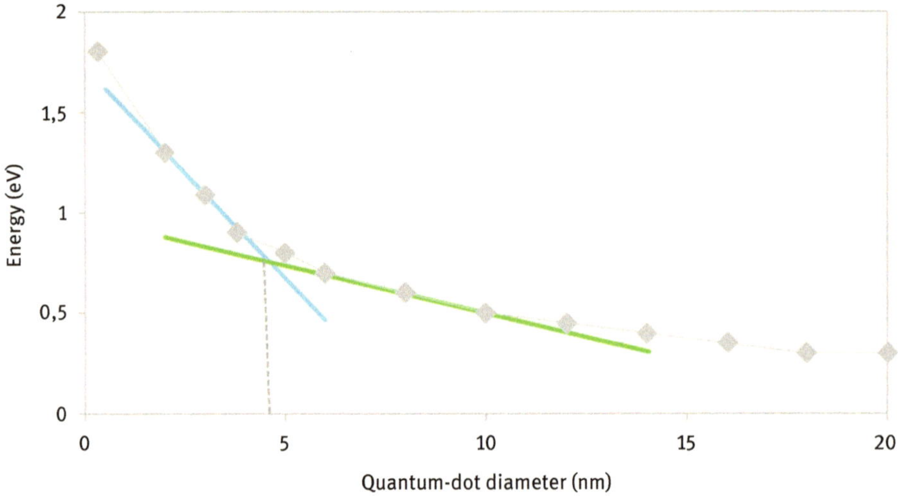

Fig. 8.3: The energy versus dimension of PbSe.

the oxidation of organics as they are more active in producing ·OH radicals. Interestingly, there is a correlation between the size of particles and their oxidative activity toward different organic molecules: phenol is better oxidized by particles having a dimension of *ca.* 25 nm [13], while chlorinated hydrocarbons prefer smaller particles (7–11 nm) [14]. This may be somehow correlated with the light absorption and light scattering phenomena other than charge separation and transfer dynamics.

Also the stability of thermal catalysts and their activity may be influenced by dimension. Mixed oxides based on CeO_2 show a quite different activity and stability (life time) when nanoparticles with size of 5–7 nm are used in the carboxylation of alcohols [15].

The nanoscale dimension raises the issue of amorphous versus crystalline form activity in several applications. Often, as mentioned above, *ca.* 5 nm may signal a quite sudden change in a property. This is shown in Figure 8.3 where the energy versus dimension of PbSe is reported.

Such correlations can be found for several systems.

8.4 Stability and toxicity

The stability of a nanoparticle means that atoms (metal nanoparticles) or ions (oxides and other compounds) are leached at a different rate from the particle. Leaching is dependent on the nature and size of the particle, and influences from one side the life of the nanoparticle (activity of the catalyst for example) and from the other the impact on biosystems. Nanoparticles may not have the same mechanism of leaching as macroparticles: as a matter of fact, the two structures have quite different surface energy.

A correlation between stability and toxicity has recently been established for several nanomaterials [16]. In some cases it has been demonstrated that coating the surface may reduce the toxicity [17], again supporting that the surface of nanoparticles plays a key role in their bio-activity. A point that is obscure is the role that aggregation of nanoparticles in a biosystem may play. Therefore, the conversion of nanoparticles when in contact with organs and cells is an issue of great importance for clarifying their potential toxicity: do nanoparticles degrade to atoms and ions or do they aggregate? Are the nanoparticles *per se* toxic or the products of their conversion? If the nanosize of particles may foster their take-up by an organism, is the nanosize responsible for the toxicity or may the size only accelerate the formation of "the toxic species" with respect to bulk materials? Several points are still obscure and need clarification for assessing the real risk attached to nanosized materials.

8.5 Fields of application of engineered nanoparticles

Nanoparticles, as mentioned above, find application in several fields, as shown in Table 8.1 that shows a perspective on the use of nanoparticles according to the EU view.

Tab. 8.1: Perspective on fields of application of nanoparticles.

Supramolecular Chemistry	Supra-molecular self-assembling	Batteries	Energy storage	Inorganic self-assembling
Materials Science	Carbon nanotubes	Nanocomposites	Nanotribology	Nanofiltration
	Nanofibers	Colorants	Suspension and inks	
	Nanomaterial coatings		Nanostructured catalysts	
Molecular Biology and Medicine	Biocatalysis	Biosensors	Biodata processing	
	DNA arrays	Biochip	Drug delivery	Nanoimplants
	Biological self-assembling		Biosynthesis of functional molecules	
Quantum Physics	Organic electronics	Photo-voltaic	Nanoelectronics	Nano-mechanics
	Magneto-resistive devices		Quantum dot LEDs and lasers	
Manufacturing	Transport	Aerospace	Cosmetics	Textiles
	Materials supply	Environmental protection	Information miniaturization	Food and beverages
	Pharmaceuticals	Electronics	Energy	Chemical

New applications are discovered year after year and the large popularity of use of nanoparticles makes more urgent the need of clarifying the risks associated with their use. Some examples of practical applications of nanosized materials are presented in Table 8.2.

Tab. 8.2: Some particular uses of nanoparticles.

Nanomaterial	Field of application	Issues	References
TiO_2	Photocatalysis (oxidation of organics)	Release	[18]
TiO_2, ZnO	Paints, cements, sunscreen, UV-protection, Batteries	Release	[19]
Carbon-nanotubes	Adsorption of contaminants, catalysis, H_2-storage, lubricants, additive to tires	Disposal	[20]
Ag^+	Antimicrobial, textiles, varnishes	Release	[21]
Pt, Pd, Rh	Automotive converters	Release as aerosols	[22]
SiO_2	Ceramics, fire-resistant glass, varnish, dentistry, pharmaceuticals	Respiratory apparatus effects	[23]
Fe_2O_3	Additive to cement	Release	[24]

In many applications nanomaterials substitute other species that may have a high toxicity, resulting in a beneficial effect.

The examples above are not an exhaustive list, but sufficiently representative of the many different products really in use. Chapter 1 of this book also lists some specific applications of nanomaterials. In some cases, they reach a large scale market (such as noble metals in converters, silver, TiO_2), others may have a quite specialist market. The large popularity of all the listed applications justifies the worries about the effects of useing nanomaterials.

8.6 Access to bio-organisms and toxicity to organisms

Nanoparticles can access a bio-organism in many different ways. During the preparation phase, they can reach the operator during the stirring phase (inhalation) of liquids, or through the gas phase, or else as particles dispersed in the atmosphere in operations such as surface cleaning, sanding, milling, mechanical disruption of bulk materials. Risk may also rise during transfer operations [18]. Ultrafine "dust" powders may not be purposely originated, and this increases the uncertainty about this category of materials. It has been proved that they can interact with macrophages and produce pulmonary pathogenesis.

The respiratory route is the most effective in taking up nanoparticles dispersed in the atmosphere. Their size means that they can reach the lungs from where they can be transferred to blood with several adverse effects [19].

Ingestion is also a possible route of access, if nanoparticles are present in medicinal drugs.

Also skin, through the derma, can be a route of up-take of nanoparticles [20]. Especially nonhealthy skin, bearing wounds or irritations, may represent a route for the easy up-take of nanosized particles. Even in the absence of damages, skin can take up nanoparticles: the mechanism is not very clear [21].

Ears and eyes can also be an access route for nanosized particles that can then accumulate or diffuse to other organs. Interestingly, nanoparticles can be delivered though such routes [22]. The accumulation of nanoparticles may cause damage.

Some nanoparticles have been shown to be toxic [23]. Accumulation in cells is well documented [24]. Only quite recently (2007, Stockholm) a forum dedicated to discussing the toxicity of nanoparticles was organized and since then the interest of academic groups in understanding the interaction of nanoparticles with cells has steadily grown [25].

The USA Environment Protection Agency has found that TiO_2 may cause brain damage in mice [26]. The risk associated with TiO_2 arises from the fact that it can induce the formation of radicals ($^{\cdot}OH$) that are very toxic: once again the property that makes TiO_2 very useful makes it quite dangerous for organisms: TiO_2 may be genotoxic and cytotoxic [27]. Fullerenes are also suspected to damage cells [28]. The cytotoxicity of nanoparticles has recently been reviewed.

The causes of the toxicity of nanosized materials has been reviewed [29] and three main causes have been identified, namely: (i) release of toxic ions; (ii) modification of cellular membranes, crossing of the membrane and interference with nuclei; (iii) piercing of cell membranes by nanoparticles with shapes of nanotubes or pillars.

The risk assessment in the environment is under continuous evolution and methodologies developed for natural nanomaterials now extend to engineered nanoparticles [30].

New tools are being developed for nanoparticle analysis that allow their classification [31].

8.7 Applications of nanoparticles in biomedicine

What is of interest is that nanoparticles find a large range of applications in biomedicine. Table 8.3 shows a list of applications. Obviously, such usage is making scientists very cautious and the effects of nanoparticles are being monitored more and more.

Tab. 8.3: Nanoparticle applications in biomedicine.

Type of nanoparticle	Application	Reference
Gold NPs	Imaging tools, phototherapy agents and gene/drug delivery carriers	[24c]
Magnetic NPs	Diagnostics and molecular biology	[32]
Magnetite and ferrofluidic NPs	Transport of chemical factors, direct drug delivery in cancer, biomanipulation	[24a, 33, 34]
Fluorescent CdSe and CdSe/ZnS NPs	Labeling of biomolecules, contrasting agents in visualization of physiological processes in magnetic resonance imaging (MRI)	[29]
Silver NPs	Antibacterial action	[35]
Silver and copper NPs	Inhibition of growth of microorganisms	[36]

8.8 Measurement of the concentration

The measurement of the concentration of nanoparticles in a workplace, namely in a pilot-scale facility at the Institute of Energy and Environmental Technology e.V. [31], has been performed using an integrated set of techniques. The measurements were carried out during the production of nanoparticles of silicon, the process of packaging, cleaning and overnight. The conclusion of the authors was that the concentration was comparable, for that particular plant, to the background measured in several other monitored processes. The authors also concluded that the protection of workers was always desirable. The protection needed is not different from that used in other plants or laboratories, such as: respiratory masks; hair, eye and hand protection. It would be interesting to have similar tests at other workplaces, such as those where nanoparticles are used.

8.9 Conclusions

Nanomaterials find applications in several areas, including biomedicine in which they are used as a delivery tool, and they are gaining more and more popularity in many different fields. This means that their risk assessment is a hot topic today. Several nanoparticles have been shown to be toxic, for other nanosized particles their effect on biosystems must be demonstrated. Studies on the interaction of nanoparticles with cells have been undertaken quite recently. Little is known about the mechanism of action, while routes that may be used for the up-take are today well clarified. More work is necessary for making clear the role of nanoparticles and the number of research groups that investigate this field is growing continuously.

Bibliography

[1] (a) Paull R, Wolfe J, Hebert P, Sinkula M. Investing in nanotechnology. Nature Biotechnol 2003; 21: 1134–1147. (b) Academie des Sciences, Academie des Technologies. Nanosciences-Nanotechnologies. Science and Technology Report No. 18 (French Academy of Sciences, 2004); summary of recommendations (in English) available at http://www.tinyurl.com/nqwdda. (c) The Royal Academy of Engineering. Nanoscience and Nanotechnology: Opportunities and Uncertainties (The Royal Society, 2004); available at http://www.Nanotech.org.uk. (d) Hansen SF, Larsen BH, Olsen SI, Baun A. Categorization framework to aid hazard identification of nanomaterials. Nanotoxicology 2007; 1: 243–250. (e) Nanoscale Science Engineering and Technology Subcommittee. The National Nanotechnology Initiative: Strategic Plan (US National Science and Technology Council, 2004; available at http://www.Nano.gov/NNI_Strategic_Plan_2004.pdf.

[2] Maynard AD. Nanotechnology: A Research Strategy for Addressing Risk. Woodrow Wilson International Center for Scholars, Washington, 2006

[3] Rejeski D, Lekas D. Nanotechnology field observations: Scouting the new industrial west. Journal of Cleaner Production 2008; 16: 1014–1017.

[4] Dawson NG. Sweating the small stuff: Environmental risk and nanotechnology. Bioscience 2008; 58(8): 690.

[5] Jolivet JP et al. Size tailoring of oxide nanoparticles by precipitation in aqueous medium. A semi-quantitative modeling. J. Mater. Chem. 2004; 14: 3281–3288.

[6] (a) Lamber R, Wetjen S, Jaeger NI. Size dependence of the lattice parameter of small palladium particles. Phys. Rev. B 1995; 51: 10968–10971. (b) Ayyub P, Palkar VR, Chattopadhyay S, Multani M. Effect of crystal size reduction on lattice symmetry and cooperative properties. Phy. Rev. B 1995; 51: 6135–6138. (c) Anastasio C, Martin ST. Atmospheric Nanoparticles, in: Banfield JF, Navrotsky A. (eds.), Nanoparticles and the Environment, pp. 293–349 Mineralogical Society of America, Washington D.C. 2001. (d) Brice-Profeta S et al. Magneticorder in γFe_2O_3 nanoparticles: A XMCD study. J. Magn. Magn. Mater. 2005; 288: 354–365. (e) Baudrin E et al. Structural evolution during the reaction of Li with nanosized rutile type TiO_2 at room temperature. Electrochem. Commun. 2007; 9: 337–342.

[7] (a) Chen Q, Zhang ZJ. Size-dependent superparamagnetic propertiesof $MgFe_2O_4$ spinel ferrite nanocrystallites. Appl. Phys. Lett. 1998; 73: 3156–3158. (b) Selbach SM, Tybell T, Einarsrud MA, Grande T. Size-dependent properties of multiferroic $BiFeO_3$manoparticles. Chem. Mater. 2007; 19: 6478–6484. (c) Shetty S, Palkar VR, Pinto R. Size effect study in magnetoelectric $BiFeO_3$ system. Pramana 2002; 58: 1027–1030. (d) Chattopadhyay S, Ayyub P, Palkar VR, Multani M. Size-induced diffuse phase transition in the nanocrystalline ferroelectric $PbTiO_3$. Phys. Rev. 1995; B52: 13177–13183. (e) Shih WY, Shih W-H, Aksay IA. Size dependence of the ferroelectric transition of small $BaTiO_3$ particles: effect of depolarization. Phys. Rev. 1994; B50: 15575–15585.

[8] (a) Dormann JL, Fiorani D, Tronc E. Magnetic relaxation in fine-particle systems. Adv. Chem. Phys. 1997; 98: 283–494. (b) Gangopadhyay S et al. Magnetic properties of ultrafine iron particles. Phys. Rev. 1992; B 45: 9778–9787. (c) Pastor GM, Dorantesdavila J, Bennemann KH. Size and structural dependence of the magnetic properties of small 3d-transition-metal clusters. Phys. Rev. 1989; B 40: 7642–7654.

[9] Zhang M et al. Size-dependent melting point depression of nanostructures: Nanocalorimetric measurements. Phys. Rev. 2000; B62: 10548–10557.

[10] Rusanov AI. Surface thermodynamics revisited. Surf. Sci. Rep. 2005; 58: 111–239.

[11] (a) Haruta M. Size- and support-dependency in the catalysis of gold. Catalys. Today 1997; 36: 153–166. (b) Sau TK, Pal A, Pal T. Size regime dependent catalysis by gold nanoparti-

cles for the reduction of eosin. J. Phys. Chem. 2001; B 105: 9266–9272. (c) Cortie MB, Van der Lingen E. Catalytic gold nanoparticles. Mater. Forum 2002; 26: 1–14. (d) Turner M et al. Selective oxidation with dioxygen by gold nanoparticle catalysts derived from 55-atom clusters. Nature 2008; 454: 981–983.

[12] (a) Dibenedetto A, Aresta M, Distaso M, et al. High throughput experiment approach to the oxidation of propene-to-propene oxide with transition-metal oxides as O-donors. Catalysis Today 2008; 137: 44–51. (b) Aresta M, Dibenedetto A, Carbon dioxide as building block for the synthesis of organic carbonates behavior of homogeneous and heterogeneous catalysts in the oxidative carboxylation of olefins. J. Mol. Catal. 2002; 182–183: 399–409.

[13] Almquist CB, Biswas P. Role of synthesis method and particle size of nanostructured TiO_2 on its photoactivity. J. Catalys. 2002; 212: 145–156.

[14] (a) Maira AJ, Yeung KL, Lee CY, Yue PL, Chan CK. Size effects in gas-phase photo-oxidation of trichloroethylene using nanometer-sized TiO_2 catalysts. J. Catalys. 2000; 192: 185–196. (b) Wang CC, Zhang Z, Ying JY. Photocatalytic decomposition of halogenated organics over nanocrystalline titania. Nanostruct. Mater. 1997; 9: 583–586.

[15] (a) Aresta M, Dibenedetto A, Pastore C, Angelini A, Aresta B, Pápai I. Influence of Al2O3 on the performance of CeO2 used as catalyst in the direct carboxylation of methanol to dimethylcarbonate and the elucidation of the reaction mechanism. Journal of Catalysis 2010; 269: 44–52. (b) Dibenedetto A, Aresta M, Angelini A, Ethiraj J, Aresta BM. Synthesis, characterization, and use of Nb V/Ce IV-mixed oxides in the direct carboxylation of ethanol by using pervaporation membranes for water removal. Chem. Eur. J. 2012; 18: 10324–10334.

[16] (a) Franklin NM, et al. Comparative toxicity of nanoparticulate ZnO, bulk ZnO, and $ZnCl_2$ to a freshwater microalga (Pseudokirchneriella subcapitata): The importance of particle solubility. Environ. Sci. Technol. 2007; 41: 8484–8490. (b) Morones JR, et al. The bactericidal effect of silver nanoparticles. Nanotechnology 2005; 16: 2346–2353.

[17] Derfus AM, Chan WCW, Bhatia SN. Probing the cytotoxicity ofsemiconductor quantum dots. Nano Lett. 2004; 4: 11–18.

[18] Lam CW, James JT, McCluskey R, Arepalli S, Hunter RL. A review of carbon nanotube toxicity and assessment of potential and environmental health risks. Crit Rev Toxicol 2006; 36: 189–217.

[19] Ballou B, Lagerholm BC, Ernst LA, Bruchez MP, Waggoner AS. Non-invasive imaging of quantum dots in mice. Bioconjugate Chem 2004; 15: 79–86.

[20] Blundell G, Henderson WJ, Price EW. Soil particles in the tissue of the foot in endemic elephantiasis of the lower legs. Ann Trop MedParasitol 1989; 83: 381–5.

[21] Navarro E, Baun A, Behra R, Hartmann NB, Filser J, Miao A-J, Quigg A, Santschi PH, Sigg L. Environmental behavior and ecotoxicity of engineered nanoparticles to algae, plants and fungi. Ecotoxicology 2008; 17: 372–387.

[22] Mamedova N, Dormer K, Kopke R, Chen K, Liu J, Ronald J, Costello M, Gibson D, Mondalek F. Feasibility of superparamagnetic NPs for drug delivery to the inner ear. 2005; http://www.nanobmi.com/images/Abstract_Nano_2005.pdf

[23] (a) Mraz SJ.Nanowaste: The next big threat? Machine Design 2005; 77: 46–53. (b) Oberdorster G, Maynard A, Donaldson et al. Principles for characterizing the potential human health effects from exposure to nanomaterials: Elements of a screening strategy. Particle and Fibre Toxicology 2005; 2: 8–43.

[24] (a) Liu WT. Nanoparticles and their biological and environmental applications. Journal of Bioscience and Bioengineering 2006; 102: 1–7. (b) Marshall AT, Haverkamp RG, Davies CE, Parsons JG, Gardea-Torresdey JL, van Agterveld D, Accumulation of gold nanoparticles in Brassicjuncea. International Journal of Phytoremediation 2007; 9: 197–206. (c) Verma A,

Uzun O, Hu Y, Hu Y, Han HS, Watson N, Chen S, Irvine DJ, Stellacci F. Surface-structure-regulated cell-membrane penetration by monolayer-protected nanoparticles. Nature Materials 2008; 7; 588–595.
[25] (a) Kostarelos K, Lacerda L, Pastori et al. Cellular uptake of functionalized carbon nanotubes is independent of functional group and cell type. Nature Nanotechnology 2007; 2: 108–113. (b) Porter AE, Gass M, Muller K, Skepper JN, Midgley PA, Welland M, Direct imaging of single-walled carbon nanotubes in cells. Nature Nanotechnology 2007; 2: 713–717.
[26] (a) Long TC, Saleh N, Tilton RD, Lowry GV, Veronesi B, Titanium dioxide (P25) produces reactive oxygen species in immortalized brain microglia (BV2). Implications for nanoparticle neurotoxicity. Environmental Science and Technology 2006; 40: 4346–4352. (b) Long TC, Tajuba J, Sama P, Saleh N, Swartz C, Parker J, Hester S, Lowry GV, Veronesi B. Nanosize titanium dioxide stimulates reactive oxygen species in brain microglia and damages neurons in vitro. Environmental Health Perspectives 2007; 115: 1631–1637.
[27] (a) Vevers WF, Jha AN, Genotoxic and cytotoxic potential of titanium dioxide (TiO_2) nanoparticles on fish cells in vitro. Ecotoxicology 2008; 17: 411–421. (b) Handy RD, Henry TB, Scown TM, Johnston BD, Tyler CR, Manufactured nanoparticles: Their uptake and effects on fish – a mechanistic analysis. Ecotoxicology 2008; 17: 396–410.
[28] Lewinski N, Colvin V, Drezek R, Cytotoxicity of nanoparticles. Small 2008; 4: 26–49.
[29] Kirchner C, Liedl T, Kudera S, Pellegrino T, Javier AM, Gaub HE, Stolzle S, Fertig N, Parak WJ. Cytotoxicity of colloidal CdSe and CdSe/ZnSnanoparticles. Nano Letters 2005; 5: 331–338.
[30] Bystrzejewska-Piotrowska G, Golimowski J, Urban PL. Nanoparticles: Their potential toxicity, waste and environmental management. Waste Management 2009; 29: 2587–2595.
[31] Wang J, Asbach C, Fissan H, et al. Emission measurement and safety assessment for the production process of silicon nanoparticles in a pilot-scale facility. J Nanopart Res 2012; 14: 759.
[32] Ngomsik AF, Bee A, Draye M, Cote G, Cabuil V, Magnetic nano- and microparticles for metal removal and environmental applications: A review. Comptes Rendus Chimie 2005; 8: 963–970.
[33] Arruebo M, Fernandez-Pacheco R, Ibarra MR, Santamaria J. Magnetic nanoparticles for drug delivery. Nano Today 2007; 2: 22–32.
[34] Shan GB, Xing JM, Zhang HY, Liu HZ. Biodesulfurization of dibenzothiophene by microbial cells coated with magnetite nanoparticles. Applied and Environmental Microbiology 2005; 71: 4497–4502.
[35] Weir E, Lawlor A, Whelan A, Regan F. The use of nanoparticles in antimicrobial materials and their characterization. Analyst 2008; 133: 835–845.
[36] Yoon KY, Byeon JH, Park JH, Hwang J. Susceptibility constants of Escherichia coli and Bacillus subtilis to silver and copper nanoparticles. Science of the Total Environment 2007; 373: 572–575.

Index

Also of Interest

Nanocarbon-Inorganic Hybrids. Next Generation Composites for Sustainable Energy Applications
Dominik Eder, Robert Schlögl (Eds.), 2013
ISBN 978-3-11-026971-0, e-ISBN 978-3-11-026986-4

Nanocomposites
Materials, Manufacturing and Engineering
Paulo J. Davim, Constantinos A. Charitidis (Eds.), 2013
ISBN 978-3-11-026644-3, e-ISBN 978-3-11-026742-6

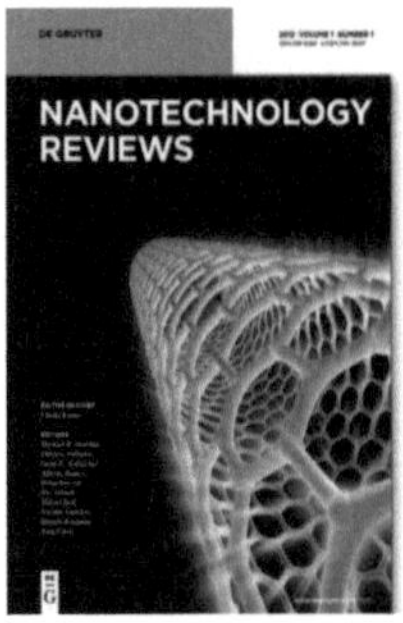

Nanotechnology Reviews
Challa Kumar (Editor-in-Chief)
ISSN 2191-9089
e-ISSN 2191-9097

Bei Fragen zur Produktsicherheit wenden Sie sich bitte an:
If you have any questions regarding product safety,
please contact:

Walter de Gruyter GmbH
Genthiner Straße 13
10785 Berlin
productsafety@degruyterbrill.com